Acclaim for
Prostitution and Trafficking in Nevada: Making the Connections

Prostitution and Trafficking in Nevada: Making the Connections is a riveting analysis of the enormous human tragedy associated with prostitution and sexual trafficking and the complex forces that shape and sustain them. Through careful scholarship and documentation, withering analyses, compelling victim commentaries, and a fearless citation of names, places, and policies, this volume must now be considered the new standard of excellence for understanding and responding to this age-old problem that ultimately victimizes us all. The author assails "legalized" prostitution as nothing more than a glamorized cover for exploitation, abuse, and violence of the thousands of women forced to work in the sex industry. All the public relations efforts to glamorize prostitution and to label it as a "victimless crime" collapse before Dr. Farley's systematic presentation of the facts and living voices behind this often hidden or disguised social problem. Her powerful account of "What happens in Vegas" -- to prostitutes and sex trafficking victims of all ages - educate and persuade the reader that prostitution destroys lives, and profits only the pimps, criminals, and corrupt public officials who collect the tainted money, and who keep the problem above the laws and moral scrutiny. This is an age-old story, but it has never been told better, nor with as much meaning and force. Must be read!

Anthony Marsella, PhD, Professor Emeritus at University of Hawaii, editor of Ethnocultural Aspects of PTSD, published by the American Psychological Association.

Melissa Farley confronts legal prostitution, not as an isolated subject, but as a phenomenon that has a devastating impact on the social, political and legal character of Nevada. She links the justifications for legal prostitution with the immeasurable destructiveness of the trafficking of women internationally, demonstrating the damage of unquestioned male power and privilege.

Mary W. Stewart Ph.D., Professor and Director of Women's Studies, University of Nevada, Reno

Finally, here's a book about sex slavery not just in distant lands but in Nevada, USA. Dr. Melissa Farley looks at sex slavery in Nevada's legal and illegal prostitution sectors and discovers there's not much difference between the two. Farley meticulously researches and reports on the women victims-- it's not a pretty picture--and issues a moving call for help. If you've wanted to know what really goes on in the sex industry in Las Vegas and nearby Nevada counties, this is an unpleasant but must read.

John Miller, Research Professor of International Affairs, George Washington University; Ambassador at Large on Modern Day Slavery 2004-2006; Senior Advisor to the Secretary of State 2002-2006; Director of the Office to Monitor and Combat Trafficking

Dr. Farley's careful research demystifies the supposed 'glamour' in prostitution that rationalizes what is in fact victimization. Holding perpetrators accountable in every jurisdiction is essential to affirming human rights in America. There is a strong link between what Farley describes and the sexual assaults/prostitution of children and youth.

Sharon W. Cooper, MD FAAP, Consultant, National Center for Missing and Exploited Children, University of North Carolina Chapel Hill School of Medicine

This book, beyond a reasonable doubt, presents factual information of what prostitution is truly about in the state of Nevada. It debunks the mis-education and advertised myths about legal and illegal prostitution, including the glamorization and glorification of the pimp and prostitute sub-culture. I investigated child and adult prostitution cases for 23 years of my 26-year career as an FBI Agent. My last 20 years were in Las Vegas, Nevada. It is clear to me that Melissa Farley has presented what all citizens need to know about prostitution, and that is …"to seek the Truth and see the total picture."

Roger Young, Special Agent, FBI, Retired, Reno Nevada

Reading Melissa Farley's outstanding book reminded me of how similar recognition of the sexual abuse of children is to the recognition of effects of pornography, also a form of abuse. In both cases, professionals have been slow to recognize the harm and have often not listened to those most immediately (and deeply) affected. One of the most valuable chapters in this remarkable work of scholarship is the account by those most directly involved of how prostitution works much like the indoctrination into a cult. The book is intellectually exhilarating, reminding us of the deep issues involved in prostitution. I highly recommend it to anyone concerned with the sexual abuse of women.

Jeffrey Masson Ph.D., author of *The Assault on Truth: Freud's Suppression of the Seduction Theory*, who has published 24 books since 1984.

Prostitution and Trafficking in Nevada: Making the Connections

Prostitution and Trafficking in Nevada: Making the Connections

Melissa Farley

Prostitution Research & Education
San Francisco, California

Prostitution and Trafficking in Nevada: Making the Connections

Prostitution Research & Education
P.O. Box 16254
San Francisco, CA 94116-0254
www.prostitutionresearch.com

ISBN-13: 978-0-6151-6205-8

**This book is dedicated to Andrea Dworkin,
beloved sister and teacher.**

**Melissa Farley's statement about the front and back cover
photographs:**

I shot the front cover photograph in June 2007 at a place named Mound House in northern Nevada. The photo is a legal brothel although at first glance people think it's a prison yard. Women inside the Nevada legal brothels describe them as little prisons in double-wide trailers. Legal prostitution feels like a prison to them.

Mound House is located where the Paiute and Shoshone people used to live. While there are many descendants of the original inhabitants of Mound House in Nevada today, it is horribly ironic that a Nevada legal brothel, which is a place that promotes the destruction of women's bodies and psyches, was built in a place where genocide was attempted in North America.

The back cover of the book has a photograph of one of the millions of advertisements for illegal prostitution in Las Vegas. Newspaper flyers and business cards like the one shown here are distributed 24 hours of the day in Las Vegas. The ads provide a welcoming environment to men who prey on women, and they persuade visitors and johns that prostitution is legal there. In fact, prostitution is illegal in Las Vegas, so this card advertises an illegal activity.

The card on the back cover describes the woman who was photographed in prostitution as a slave girl, offering her up to the john. Whether this description of her as a slave girl is intended literally or not, the truth is that women and children in prostitution feel enslaved, and they are enslaved by multiple forms of visible and invisible coercion into prostitution, by lethal sexist devaluation of women, by life-threatening poverty, by race/ethnic hatred, and by prior sexual, physical and emotional abuse that are so damaging that prostitution appears to be a way out. Tragically, prostitution is not a way out. Witnesses and survivors the world over, including witnesses and survivors of Nevada legal prostitution, view prostitution and trafficking as a form of modern-day slavery.

Kathy Watkins designed both the front and back covers of the book.

CONTENTS

Acknowledgments

This book was produced with support from Prostitution Research & Education, a nonprofit organization in San Francisco, California, and also with support from Grant #2074-610001 from the Office to Monitor and Combat Trafficking in Persons (TIP) of the United States Department of State. The TIP Office provided the funding to research prostitution and trafficking in Nevada and to produce a report for them. This support made possible the original research that formed the basis for this book but it has not affected the book's content for which the author is solely responsible.

The funding from the TIP Office was based on United States National Security Policy Directive 22, which reads in part:

> Our policy is based on an abolitionist approach to trafficking in persons, and our efforts must involve a comprehensive attack on such trafficking, which is a modern day form of slavery. In this regard, the United States Government opposes prostitution and any related activities, including pimping, pandering, or maintaining brothels as contributing to the phenomenon of trafficking in persons. These activities are inherently harmful and dehumanizing. The United States Government's position is that these activities should not be regulated as a legitimate form of work for any human being.

I am grateful to Ambassador John Miller formerly of the Trafficking in Persons Office. This book came into being because of his vision and leadership.

Catharine A. MacKinnon at the University of Michigan Law School has been my mentor in a relationship that is far outside the academy and far more generous than any traditional teacher. While most know her as a feminist legal scholar and expert on constitutional and international law, she also has a profound awareness of traumatic stress, dissociation, and the psychological destruction caused by sexual violence throughout women's lifetimes - from incest to rape to prostitution/trafficking to genocide by rape. Catharine MacKinnon's editing, critical feedback, and her humor were important in the creation of this book.

One of the first people I met when I began the two and half-year process of interviewing people who knew about prostitution and trafficking in Nevada was Terri Miller, who at that time was Education Coordinator of the Nevada Coalition Against Sexual Violence. Her generosity in sharing what she knew, which included extensive information about Nevada agencies and advocacy groups, was vital to the startup of this project. I consulted with her throughout the years I was conducting this research, and she was constantly available and supportive. In 2007, Terri Miller was Program Director of the Anti Trafficking League Against Slavery of the Las Vegas Metropolitan Police Department.

Steve Miller, a former Las Vegas City Councilman now an intrepid writer on the topics of disorganized and organized crime in Las Vegas, has provided me and others with information on organized crime that no one else dares to know about (http://www.stevemiller4lasvegas.com/). As Judi McLeod wrote about him: "No mobster boss, no matter how threatening, no politician no matter how corrupt or influential, no lawyer gone wrong, is safe from his pen." (Judi McLeod, Canada Free Press, March 26, 2007. http://www.canadafreepress.com/2007/judi032607.htm).

Kathy Watkins has made so many contributions to this book that it's difficult to list them all. She conducted the research and produced the data that formed the basis for the chapter on advertising, and she also worked with me to investigate several of the most challenging topics in the book. She is a book publisher and she typeset the book, indexed it, and designed the front and back cover. Her passion for justice for prostituted and trafficked women drove her work on this project and sustained me as well.

My dear friends Wendy Freed and Harvey Schwartz offered me the emotional support to do this work. I wish that other researchers of sexual violence had access to the loving and insightful support provided by these therapists who are on the Prostitution Research & Education board of directors.

Democratic Congresswoman from New York Carolyn Maloney generously agreed to write a Foreword to the book. She has served in the U.S. House of Representatives since 1993 and is currently Co-Chair of the Congressional Caucus on Human Trafficking. Congresswoman Maloney has authored and passed legislation that targets the demand side of sex trafficking, and was pivotal in the passage of the landmark 2007 New York antitrafficking law.

Jebbie Whiteside J.D. was a law student at University of Nevada, Las Vegas when she first contacted me about her paper on legal brothel prostitution and domestic violence. She worked with me in locating the currently operating brothels, and in conducting interviews of the women in the legal brothels. Her northern Nevada easy-going manner and her sense of humor helped defuse many challenging situations.

Feminist comedian Betsy Salkind is a source of creative inspiration and fall-off-your-chair-humor about topics no one else dares to laugh about. She never fails to come up with a devastatingly on-point response to sexual violence against children and adults. She made major contributions to the cartoons. Feminist cartoonist bulbul who has drawn cartoons about the labor movement and the women's movement for 30 years, drew the cartoons in this book.

I have found a community of like-minded people in Nevada. Although this book is about the pimps, johns, traffickers and those complicit with them by their silence – there are many many others in Nevada who care deeply about the sexual violence of prostitution/trafficking and are working for change. Some of these people include Ofelia Monje who was Bilingual Victim Advocate at the Las Vegas Metropolitan Police Department when I spoke with her; Jodi Tyson, former director of the Nevada Coalition against Sexual Violence and currently Youth Suicide Prevention Program Coordinator, Nevada Office of Suicide Prevention, Department of Health and Human Services in Las Vegas. Jodi Tyson is a member of the Nevada community whose work is about violence prevention; she helped me to understand the intrinsically harmful nature of Nevada prostitution, even if legal; Glen Meek, veteran investigative reporter who has broken many stories on prostitution and trafficking at KTNV-TV, Las Vegas; Aaron Stanton, Vice Detective in the Las Vegas Metropolitan Police Department who was tolerant of my ignorance about Nevada prostitution/trafficking when I started out and generously provided me with his time, information, and support, as did Detective Don Fieselman, Vice Section Las Vegas Metropolitan Police Department; Lieutenant David Logue, Las Vegas Metropolitan Police Department, Intelligence Division; Howard Meadow, Alternative Sentencing and Education Division of the Las Vegas Municipal Court; Candice Trummell, Chair of the Nye County Commission, Pahrump who is the kind of politician I want in my home town; Amber Batchelder, Director of SafeNest Domestic Violence Shelter, Las Vegas; Sergeant Dave Evans, Reno Police Department; Sergeant Gil Shannon, Las Vegas Metro Police Department; Lieutenant Brad Simpson, Detective Bureau, Las Vegas Metropolitan Police Department; Todd Palmer, who was with the FBI in Las Vegas when I spoke with him; Roger Young, a retired Nevada FBI agent who now consults with community groups about the role of the media and the Internet in promoting violence against women; Captain Terry Lesney who was with the Crimes Against Youth/Family Bureau of Las Vegas Metropolitan Police Department when I spoke with her; Liza Conroy, Assistant City Attorney, Henderson; Char Hoerth, Sex Offender Registry Supervisor, Records and Identification Bureau, State Health Division, Carson City; Kevin Morss, Youth Outreach Coordinator at Westcare, Las Vegas; Miranda Smith who was Prevention Education Coordinator, Nevada Coalition Against Sexual Violence, Henderson; Lyn Amie, Operations Manager, Las Vegas Rape Crisis Center; John L. Smith, author and columnist at Las Vegas Review-Journal; Kathy Jacobs, Executive Director of the Crisis Call Center, Reno; Quentin Heiden at the U.S. Department of Labor, Las Vegas; Shizue Hill, Counseling Coordinator at Las Vegas Rape Crisis Center; Christina Hernandez, Community Outreach/Education Coordinator, Las Vegas Rape Crisis Center; Kareen Prentice, Health Information Manager, Nevada Public Health Foundation, Carson City; Tenna Herman, Management Analyst, Records and Identification Bureau, Nevada Department of Public Safety, Carson City; Lynette Smith at the Records and Identification Bureau, Nevada Department of Public Safety. Carson City; Traci Dory, Victim Services Officer, State of Nevada Department of Corrections, Carson City; Wei Yang, M.D., Ph.D. who was State Chief Biostatistician and Director, Center for Health Data and Research, Carson City when he spoke with me; Kelly Langdon, Rape Prevention Coordinator, Department of Human Resources, State Health Division, Carson City; Andrea Rivers, Department of Human Resources, State Health Division, Carson City; Tom Petersen, The Salvation Army, Reno; Karen Gedney, M.D. Senior Physician, Nevada Department of Corrections, Carson City; Maureen Budahl, LVN, public health nurse in Nye County.

Because the reach of legal and illegal prostitution in Nevada and the trafficking of women and children into and out of the state - many colleagues outside Nevada helped me in the research for this book.

They include Rachel Durchslag at the Chicago Coalition Against Sexual Exploitation who made valuable contributions to the data entry and analysis of the interviews of the women in the legal brothels; Ann Cotton, PsyD, who is at the Department of Psychiatry and Behavioral Sciences, University of Washington School of Medicine, Seattle, ran the statistical analyses of the data on both the women in the legal brothels and the men at University of Nevada, Reno. Dr. Cotton's expertise in the area of sexual violence and prostitution was invaluable in this work. Global trafficking expert Dr. Janice Raymond of the Coalition Against Trafficking in Women has provided me with invaluable consultation and advice over the past decade. Dr. Eleanor Kennelly Gaetan at the Office to Monitor and Combat Trafficking in Persons of the U.S. State Department played a vital role in implementing this research study from start to finish; Jessica Neuwirth at Equality Now in New York has also offered me invaluable strategic advice on the use of this research to promote equality for women and girls; Bradley Myles, Polaris Project, Washington, D.C.; Lisa Thompson, Salvation Army, Alexandria, Virginia; Detective Constable Raymond Payette was at Youth Squad-KEYS Vancouver Police Department, Vancouver, Canada when I interviewed him and has been a consistent source of information about trafficking into North America; Annie Fukushima, a student at University of California, Berkeley researched some of the legal cases presented here, has written about racism in telephone directory advertising for prostitution in the United States; Paddy Lazar, former staff of Council for Prostitution Alternatives, Portland Oregon; Officer Marsha Carson, Vice Division, Portland Oregon Bureau of Police; Chong N. Kim of MAISIE (Minorities & Survivors Improving Empowerment) in Minneapolis provided valuable information about prostitution and trafficking in Nevada;

Katrina Kayden, J.D. at Villanova University who researched the Nevada state and county laws on prostitution and legal case decisions on prostitution and trafficking; Victoria Brescoll, Ph.D. at Yale University who researched the newspaper coverage and public opinion about Nevada prostitution; Lois Lee, Children of the Night, Los Angeles; Adam Freer, Salvation Army; Hayang Kim who translated the cartoon about Korean massage brothels into Korean translation; Scott Wiseman provided valuable editing of the book; Phil Williams at the Ridgeway Center for International and Security Studies, University of Pittsburg and Wade Ewing at the Ridgeway Center were generous with the information they provided me about international organized crime and trafficking.

There were many other survivors of prostitution, their friends and family, and concerned community members who because of risk to themselves or their families do not want to be named.

Foreword
Congresswoman Carolyn Maloney

Once upon a time, there was the naive belief that legalized prostitution would improve life for prostitutes, eliminate prostitution in areas where it remained illegal and remove organized crime from the business. As founder and Co-Chair of the Congressional Human Trafficking Caucus, I know that like all fairy tales, this turns out to be sheer fantasy.

As this book clearly shows, rather than improving conditions for women in prostitution, Nevada's legalized prostitution has improved conditions for pimps and brothel owners. The women are often found living in prison-like conditions, locked in or forbidden to leave the brothel. They service large numbers of customers at great risk to their own health. Ironically, the prostitutes are required to undergo periodic health checks, but the customers never have to undergo any health checks. The legal brothel owners and pimps typically pocket half of the women's earnings just like the illegal pimps do. Meanwhile, the women must pay tips and other fees from their meager earnings to the owners and staff of the brothel, as well as finders' fees to the cabdrivers who bring the clients. Women work 12 or 14 hours days and do not have veto power over their clients. Further, since communities are ashamed of the prostitutes in their midst, the women remain pariahs in the society.

The myth that legalizing prostitution and limiting it to certain counties will eliminate prostitution elsewhere in the state has proven equally false. The sex trade flourishes in areas where prostitution is illegal, like Las Vegas and Reno. Further, legal prostitution creates a culture of prostitution in the state. Aware of Nevada's lax view, many customers do not even realize that they are violating the law when they patronize prostitutes in Las Vegas and other counties. Escort services offering sexual services are ubiquitous, apparently, with over 170 pages of the Las Vegas yellow pages devoted to "entertainers" and "massage."

Organized crime likely remains a significant feature of the business. Many women engaged in legal and illegal prostitution in Nevada have been trafficked into the country or from elsewhere in the United States. The people involved in trafficking women into the Nevada sex businesses - both legal and illegal - have ties to organized crime.

Unfortunately, Nevada has become one of the main destinations for victims of human trafficking, making the state a significant player in 21st Century slavery. According to Las Vegas's Eyewitness News Channel 8, in a program aired April 24, 2007, Las Vegas is "one of the worst cities in the world for sex trafficking."[1]

In January 2007, the U.S. Department of Justice named Las Vegas as one of the 17 most likely destinations for sex trafficking victims.[2] According to Detective Constable Raymond Payette of the Vancouver Canada police department's vice squad, Las Vegas is the "epicenter of North American prostitution and trafficking." Despite its illegality, the sex industry in Las Vegas reportedly generates somewhere in the neighborhood of one to six billion dollars a year.

Human trafficking is a growing enterprise that is reaping huge profits around the globe. The United Nations says human trafficking is now the world's third largest criminal enterprise, behind only drugs and arms dealing. Traffickers earn vast amounts of money by exploiting their victims. The Polaris Project estimates that a street pimp controlling four women can earn over $630,000 per year.[3] Owners of legal brothels owners are likely to earn similar or greater amounts via the legal sexual exploitation of women in prostitution.

Reportedly, more than 600,000 - 800,000 people are trafficked across international borders for labor and commercial sex purposes each year. Some 80 percent of those people are female, nearly half are minors and most are trafficked into the sex industry. Millions of additional victims are trafficked within country.[4]

The lives of trafficking victims are pure horror. Many are tricked into entering the United States, fooled into believing that they will be doing legitimate jobs. They arrive, many with limited English skills, and have everything taken from them. Their documents, if they have any, are held by the trafficker. They see very little of the money they earn. Cut off from the outside world, they usually have no freedom of movement and no friends or relatives to help them. As you read this book, it will be clear that the experiences of trafficking victims are paralleled by the experiences of women working in Nevada's legal brothels.

A woman spoke in Reno about her experience in a legal brothel saying, "there were was little freedom but plenty of fear working at the brothel. If you want to get down to brass tacks, this was slavery, but it was better than being on the streets."[5]

Under federal law, a child under 18 years who is commercially sexually abused is a victim of trafficking. Unfortunately, child prostitutes are usually arrested by local law enforcement officials who treat them as criminals instead of victims who should be directed to agencies who can treat them. A colleague of mine who was a judge once told me that she had sentenced hundreds of prostitutes during her career. One day she realized that most of the defendants before her were really teenagers. It occurred to her that somewhere in the background was an adult who was pimping these girls, and that the johns were all adults. She wondered, why were the children being charged with a crime, while the adults went free? We went on to write a law together to address the demand side of sex trafficking.

The United States is considered both a source and a destination country. That means that we have large numbers of trafficking victims coming into the country, 14,500-17,500 per year. The victims come from East Asia, Eastern Europe, Mexico and Central America. Additionally, many young American girls and women become trafficking victims. Law enforcement sources suggest that Nevada reflects this reality, with women being trafficked into Nevada from all over the world, as well as from elsewhere in the country.

Trafficking victims often report being beaten, drugged, locked into closets or basements and otherwise abused. Even if physical abuse is not present, victims are controlled by threats against themselves and their families, as discussed in this book. In addition to a variety of different kinds of mental and emotional manipulation, victims fear of being reported to immigration authorities or threats of renewed violence. In April 2005 I was senior Democrat on the Domestic and International Monetary Policy, Trade and Technology Subcommittee. At a hearing held by the subcommittee, Tina Frundt of the Polaris Project reported that on an average night, a young woman trafficked into prosecution is forced to have sex with 10-15 johns. Women are often moved from state to state so that they cannot develop connections, and their freedom of movement is restricted.

Tina's own story is extraordinarily harrowing and is typical of women in prostitution. Sold by her foster mother's boyfriend at the age of 10, she found that no one responsible for protecting her would believe her. At 14, she ran away from the abuse at home only to be pimped out by another man. She eventually escaped, returned to her adopted family and became an outreach coordinator to help other girls like herself. When I first spoke with her, I literally could not bear to listen to her story. I kept interrupting her, not quite able to comprehend what she was telling me. When it comes to prostitution, we have all gotten into the habit of looking away, of failing to comprehend what is really happening.

I first became aware of the issue of sex trafficking in 1999, when I learned that a company in my district, Big Apple Oriental Tours, was organizing tours for men who wanted to patronize prostitutes in other countries. The owners advertised online and with brochures offering: "Introductions to lady companions throughout your stay as desired." Along with Equality Now and Gloria Steinem, I called on Janet Reno, then U.S. Attorney General, and the Queens district attorney, Richard Brown, to prosecute the company for engaging in sex tours.

In addition to advertising the tours, Big Apple Oriental Tours' brochure promised to arrange for women to immigrate to the United States "at substantial savings over the legal fees charged at home." I do not know exactly how many women came into the country through this company's efforts, but I would be willing to bet that most would properly be counted as trafficking victims.

While Big Apple Oriental Tours were clearly engaging in sex tourism, at the time, New York and federal law made it very hard to prosecute. Eliot Spitzer, who was then Attorney General of the State of New York, did bring a case against them, and obtained an injunction that temporarily shut them down. He subsequently brought a criminal case, which is still pending, and meanwhile, Big Apple Oriental Tours appears to have remained out of business.

A growing number of states are recognizing trafficking as a crime, and that is very good news. We have a responsibility to prosecute the people responsible for selling people into slavery. By the end of 2006, 27 states had passed criminal anti-trafficking legislation. Since then, both New York and Nevada have passed statutes. Of course, not all anti-trafficking statutes are equal. Nevada's simply gives prosecutors the possibility of charging traffickers with a crime.

In 1999, sex trafficking was just beginning to be recognized as a serious human rights issue. Most people simply were not aware that, in their own neighborhoods, human beings were being bought and sold. We are still in the process of learning how best to address the issue of trafficking and prostitution, and a growing number of people are joining the effort. In 2000, Congress passed the Trafficking Victims Protection Act (TVPA), which is part of the federal effort to combat sex trafficking. The law attacks the problem through prevention, prosecution and protection programs for victims.

I am very pleased that the 2005 re-authorization of TVPA (PL 109-164) includes key provisions of the End Demand for Sex Trafficking Act (End Demand Act) which I co-authored with Congresswoman Deborah Pryce (Republican, Ohio). The End Demand Act was created from the idea that we cannot end sex trafficking until we get serious about addressing the demand side of trafficking. For generations, women engaged in prostitution have been considered criminals, while the men who purchased sex from them went home to their families with impunity. It is time for these men to realize that they are the driving force behind a $10 billion criminal business that creates untold misery for its victims. The act supports the development of more effective means of targeting demand; protecting children from predators who use them in commercial sex activities; and assists State and local governments in their enforcement of existing laws dealing with commercial sex activities. It also authorized funding for a biennial survey to be conducted by the Department of Justice to determine the extent of the trafficking problem in the United States. We have the State Department's annual survey on the extent of trafficking abroad. As one of the chief destination points for traffickers, we need the statistics that show what's going on here at home.

In December 2006, the Las Vegas Metropolitan Police Department opened a new office, the Anti-Trafficking League Against Slavery (ATLAS), to combat human trafficking in the Las Vegas Valley. The task force includes the Las Vegas Metropolitan Police Department, the U.S. attorney's office, state and county prosecutors, and the Salvation Army and is funded, at least in part, through grants from the U.S. Department of Justice (DOJ). DOJ gave $370,000 to the Police Department for law enforcement efforts and $450,000 to the Salvation Army to provide social services to trafficking victims.[6]

The new ATLAS program may start to change the way law enforcement approaches the issue of prostitution. Instead of just going after pimps and others allegedly involved in prostitution rings, police crackdowns could lead to the arrests of those who do business with the underground outfits, such as the cabbies, valets and others who refer business to both the legal and the illegal brothels in the region. At the same time, investigators have the opportunity extend a helping hand to the victims of pimps and traffickers, the women in prostitution.

Law enforcement cannot rely on the federal government to address the issue of trafficking. With only 12,156 special agents for the entire nation Federal Bureau of Investigation simply lacks the manpower and resources to address the problem fully. By comparison, the New York Police Department has 37,838 officers. That is why we also need strong state laws. My state, New York, just enacted a law, under the leadership of now Governor Spitzer, that includes meaningful penalties for those who engage in trafficking, a strong definition of trafficking, and provides services for victims of trafficking.[7]

Nevada also passed a trafficking statute this year, A383, which was introduced by Assemblywoman Marilyn Kirkpatrick who represents portions of Clark County. The new measure would classify trafficking as a felony and could impose significant prison sentences on those convicted. It was approved by Nevada Governor Jim Gibbons on June 2, 2007 and will become effective on October 1, 2007.[8] While the bill gives law enforcement a new tool to address human trafficking, it unfortunately limits the violation to trafficking in illegal aliens (perversely providing no prohibition against trafficking in U.S. citizens or individuals who are legally present in the country), and does not provide funds for comprehensive victim services. It also fails to address the demand side of prostitution or sex tourism, or provide a defense for victims of sex trafficking.

In my view, it is extremely important for every state to pass trafficking legislation. A good bill would include the following:

1. **Meaningful penalties for traffickers:** Trafficking should be a felony, and should carry with it the threat of serious jail time.

2. **Meaningful penalties for buyers**. The state must address the demand for prostitution by raising penalties on those who patronize prostitution, which includes trafficking victims.

3. **A comprehensive definition of trafficking.** Too often, statutes include requirements such as imminent threat of force, which makes it much harder to prove the crime of trafficking. Traffickers use fraud, intimidation, immigration abuse, and a wide range of techniques to control their victims – fear of imminent force isn't always present, and it shouldn't be a required element to prove the crime.

4. **Sex tourism.** Sex tour businesses should not be permitted to operate in a state, regardless of the law in the destination country or state.

5. **Defense for victims.** Unless a person's status as a trafficking victim is a defense to prostitution charges, trafficking victims will convict themselves by claiming to have been trafficked.

6. **Services for Victims.** Survivors of human trafficking need a comprehensive set of services to assist them in rehabilitating their lives. They need shelter, immigration and social services, medical treatment and job training.

We will never address the issue of trafficking unless we have a comprehensive strategy that includes prosecution of traffickers, assistance for victims and strong penalties for people who make use of trafficking victims. As U.S. State Department's Ambassador Mark Lagon, Senior Advisor on Trafficking in Persons, stated: "...[W]hat we found here at home is that having a victim-centered approach -- so that people are not treated like illegal aliens or criminals, but in fact victims with rights – [is] essential."[9] The New York law addresses each of these concerns. The Nevada statute does not. Nonetheless, we can hope that programs like ATLAS will ensure that victims of sex trafficking will be treated with compassion, and traffickers will face justice.

Dr. Farley's book provides a stark window on the world of prostitution in Nevada and raises crucial questions about the advisability of legalizing prostitution. Because Nevada is a center of the sex industry in the United States, the way Nevada addresses the issue of prostitution and trafficking can have a significant impact on the sex trade. The bottom line is that we will not be able to address the issue of sex trafficking until we have a broader understanding of the industry and the women involved in the industry.

1
Introduction

In 2005, Ambassador John Miller of the Office to Monitor and Combat Trafficking in Persons of the U.S. State Department, spoke to me about his work. Miller is a man of passionate conviction who spent his time in office focused on exposing and diminishing what he called the "modern-day slavery" of trafficking in persons. Miller understood that trafficking for the purpose of prostitution is by far the most common type of trafficking in the world. According to a 2006 report, 87% of the global trafficking in persons is sex trafficking.[10]

When people in the United States first hear about trafficking, they think of someone else's country: Cambodia, Brazil, Ukraine, India. In fact, Miller spent much of his time traveling to those places in the world where there was evidence that people were being sold as sex market commodities to men in their own countries or trafficked for sale as sex to men in wealthier countries. Miller understood that prostitution was intimately connected with trafficking in persons, noting that "in addition to being inherently harmful and dehumanizing, prostitution and related activities fuel the modern-day slavery known as sex trafficking."[11]

During the course of these visits, Miller spoke with government officials and gave public speeches. Inevitably, he told me, the first question after his talk would be, "What about Nevada? Don't you have legal prostitution there?" The profound contradiction of U.S. government policies that oppose sex trafficking but which ignore or tolerate prostitution was not lost on him. He told me that his department was seeking proposals for a study on Nevada prostitution since little was known about trafficking from other countries into Nevada, trafficking from other U.S. states into Nevada, and little was known about the situation of the women in the Nevada legal brothels. He wanted to better understand the connections between legal and illegal prostitution in Nevada, the extent of organized crime, and in particular whether any prostitution in Nevada was related to trafficking in persons. I was eager to work on this project when Prostitution

Research & Education, the nonprofit organization I am affiliated with, was offered the job.[12]

Prostitution in Nevada happens in many different locations: legal brothels, illegal brothels, strip clubs, bars, hotels, massage parlors, streets, cars and private homes. We investigated the trafficking of women from a number of US states and from other countries into a range of Nevada sex businesses. It's challenging to keep up with the hydra-headed sex businesses which are constantly expanding and morphing into new entities. There are many forces that keep these global sex businesses running smoothly: poverty, sex discrimination, race discrimination, childhood trauma that creates a vulnerability to sexual predation, psychological symptoms of traumatic stress that silence victims' voices, men's attitudes of entitlement to prostitution, political corruption, organized crime, immigration policy, flawed laws and inconsistent and discriminatory enforcement of the laws against solicitation of prostitution, lack of exit services, and massive advertising for prostitution. Sometimes it seems as if there are blindfolded investigators reporting on the sex businesses' elephant in the room: one reports on the tail, another the trunk, another the ears, and these don't seem like parts of one giant beast. I wanted to better understand how the system works, and what connects it all. This is a two-year hard look at Nevada's sex industries and what keeps them running. We have every reason to assume that since the industries are part of a global market, these forces function similarly elsewhere in the world.

As the prime minister of New Zealand put it: even if prostitution is an "abhorrent" institution, maybe legalization would make it "just a little bit better."[13] We all somehow hope – against all the evidence – that there's something we can do to make prostitution immediately better for the women. That we can magically make the women safer from everything we know that happens to them but which few people want to really think about.

For the recipient, prostitution is a traumatic act of sexually exploitive predation. No matter how many lies are used to cover it up, prostitution is sexual predation, plain and simple. The existence of prostitution anywhere is society's betrayal of women and its betrayal of those who are marginalized and therefore vulnerable because of their ethnicity or poverty or abuse and neglect. The most common response to this kind of traumatic betrayal is disbelief.   When any of us who know about prostitution discuss it with the less informed, their response is often insistent disbelief.   And soon on the heels of disbelief - denial.   How does the response of shock and disbelief morph into denial?   The answer to this crucial question is that perpetrators promote the denial.[14]   Perpetrators reinforce disbelief, and use misleading words to disguise what they are actually doing to harm others. Prostitution described in fundamentalist cultures as 'short-term marriage' is such a denial.   Polygamous cult leaders' use of the term 'marriage' is a denial of socially promoted pedophilia.[15] Playboy magazine's use of the word 'hobbyist' instead of predator/john is another denial.   In the Las Vegas Yellow pages, the words 'entertainer' and 'erotic massage' are more of this same denial of the harmful nature of prostitution.

An Australian law on prostitution was changed in 1984 when it was widely assumed that legalization would decrease the damage of prostitution. Australians in the state of Victoria assumed (in error, as it turned out) that legal prostitution would contain and limit illegal prostitution, that legal prostitution would eliminate "criminal elements" associated with sex businesses, and that legalizing it would reduce the physical and sexual violence known to be connected with prostitution.[16]   Despite a lack of evidence, these same arguments are used today by proponents of legal prostitution in many parts of the world.

Nevada prostitution isn't what it looks like on HBO's Cathouse. Piercing the bubble of disbelief surrounding the Nevada sex industry, a brothel prostitution survivor said that legal brothel prostitution has a "shell around it; people look at it and say, this is a nice place, but it's really not." Another woman explained, "The Nevada sex industry is more predatory, more organized, and more international than ever. No one who makes it out ever feels safe. It really feels like a war zone, it's very survival-ish."

At first I didn't understand what she meant.   The atmosphere of overwhelming denial in Nevada negates or trivializes what is in plain sight.

The contemptuous incredulity of people with money and power has a silencing effect and makes other people question their own senses. Running deep underneath the denial and incredulity was fear. People occasionally said things like, "What I said is true but you can't quote me on that. I have to live here, [pimp's name] lives near here, my kids go to school here. I don't want my wife to get raped or killed." It was at times a struggle to be an ethical witness to the exploitation and violence against women in Nevada prostitution.

In Nevada, legal pimps are often referred to as "brothel owners." Since my goal is to lift the veil of secrecy and denial about prostitution in Nevada, I describe any person who makes money from the earnings of a woman in prostitution as a pimp. That's the legal definition of pimping. It's then easy to see how quickly the term expands to include lots of people in Nevada: legal brothel owners, illegal brothel owners, strip club owners and managers, husbands, boyfriends, yes the men lurking on streetcorners guarding their girls but also well-heeled casino hosts, taxi drivers, limo drivers, bartenders, valets, front desk managers, restauranteurs, travel agents,

Any remnants of my own denial about the dangers of prostitution in Nevada ended on the day I drove up to a legal brothel outside of Las Vegas. With a research associate, I walked into the brothel and asked to speak to the manager who under Nevada law was a legal pimp. He was used to being in charge. I was polite. As he condescendingly explained what a satisfying and lucrative business prostitution was for his "ladies," I tried to keep my facial muscles expressionless but I didn't succeed. He saw something in my eyes that he didn't like. Maybe it was skepticism or resistance to what he was saying. He whipped a revolver out of his waistband and aimed it at my head. With gun in hand, he sneered, "You know nothing about Nevada prostitution, lady. You don't even know whether or not I'll kill you in the next five minutes. That's how little you know." He also bragged about his connections, via marriage, to organized crime.

After what seemed like a very long minute, having completed his display/threat, he put the gun away, and permitted us to speak with the women in his brothel. They whispered. They peeked around corners. They were very frightened women who had no access at all to physical protection or emotional support. They had no rental car, no research associate, and no hotel with a door that locked. In fact, they weren't even allowed to have cars or car keys while at the brothel. It was a pimp's straightforward means of control: do what I say or you can't leave here. This was one of a number of experiences that enabled me to better understand the predatory, survival-ish nature of prostitution and trafficking in Nevada. Most important were interviews with survivors in and outside the legal brothels, and in illegal prostitution. I learned a great deal from interviews with pimps, with community service providers and advocates, law enforcement personnel, writers and journalists and from my collaborations with Jody Williams, Mary Stewart, and Harvey Schwartz.

Jody Williams is a survivor of prostitution who has dedicated her life to helping women escape prostitution. She founded Las Vegas-based Sex Workers Anonymous, a 12-step-oriented group for men and women escaping prostitution. Williams made important contributions to this book. She wrote Chapter 12 (Barriers to Services for Women Escaping Nevada Prostitution and Trafficking) and co-authored Chapter 3 (Pimp Subjugation of Women in Prostitution by Mind Control).

Mary Stewart, Chair of Women's Studies at the University of Nevada, Reno, also understands prostitution as a form of violence against women. She has been quoted in the Nevada press on the topic, which is how I was very fortunate to meet her. Dr. Stewart coauthored Chapter 13 which is a research study of the attitudes toward prostitution and sexually coercive behaviors of college-aged young men in Nevada. This challenging research project was only possible because of Dr. Stewart's academic commitment to understanding the ways in which sexism and racism interact to harm women. Her grasp of the far-reaching influence of Nevada's prostitution culture on all who live in the state is a crucial element of the complex issues dealt with in this book.

Harvey Schwartz is an international expert on the psychological healing of people with dissociative disorders[17] who have experienced mind control, and is the author of *Dialogues with Forgotten Voices: Relational Perspectives on Child Abuse Trauma and the Treatment of Severe Dissociative Disorders.* His contributions were the perfect complement to Jody Williams' narrative about the pimps' subjugation of women in prostitution by mind control.

I learned that the indoor pimps who control the Nevada brothels are often antisocial, unpredictable people. The popular myth that legalizing prostitution removes organized (or disorganized) criminals from the business is just that: a myth. The periodic violence against the women and the control exercised by johns and pimps in *legal* prostitution was very much like pimps' and johns' control over women in *illegal* prostitution. In the Nevada legal brothels, the women who had been there on average of 7 years were institutionalized, disconnected, and depressed.

I'd spent 13 years studying prostitution in 10 countries[18] and I predicted that prostitution in Nevada would be like prostitution elsewhere. Women in legal prostitution from Australia, New Zealand, Germany and the Netherlands experience the same physical and emotional violence as other women do in countries where prostitution is illegal. Whether prostitution is legal or illegal, women in prostitution want to get out of that life and have a safe home, a job that pays enough to live, and receive adequate medical care.

Nevada's legal brothel system makes use of "the legacy of other brothel legislation and regulation,"[19] from the Netherlands, Australia, and Germany. A Canadian researcher described Nevada as the "state pimp model."[20] State and county laws on prostitution in Nevada attempt to regulate prostitution so that it won't be too prevalent, too blatant, or too defiant.[21] Under Nevada's regulatory system, the pimp/prostitute relationship is legally redefined and enforced.[22] The only pimps who are legal are those brothel owners who have direct links with the local government.[23] A Nevada brothel-owning-pimp explained the philosophy behind Nevada prostitution: "the police chief, the City Council, the Sheriff want it to exist; they also want it hidden. If you don't get along with them, your life will be very difficult." This philosophic underpinning – keep prostitution in place but out of sight – was also present in the legal brothel system of the state of Victoria, Australia, where much of the community was opposed to the *visibility* of prostitution but supported the *existence* of prostitution for their men.[24] Victorian Australians seemed to assume the inevitability of prostitution.

The state of Nevada seems to have been born with the "stubborn idea"[25] that if a man wants sex, he has the absolute right to buy it. When its first Legislative Assembly met in 1861, Nevada's towns were all-male silver- and gold-rush camps where the only women were those who were prostituted. After the gold rush, the miners sent for their wives, but prostitution never disappeared. In 2006, there was a resurgence of barite, gypsum, and copper mining in the north of the state. Today, 87% of the gold produced in the United States is from Nevada, which is the third largest gold producer in the world, after South Africa and Australia. The same colonial practice of supplying workers with prostitutes continues today.[26]

In 1881, Nevada passed a bill to give any county commission the power to "license, tax, regulate, or suppress all houses of ill-fame."[27] In 1971, alarmed by a pimp's attempts to run a brothel in Las Vegas,[28] the Nevada Legislature prohibited county commissioners from licensing a brothel in counties with populations greater than 200,000.[29] As the population of the state grew, this prohibition was revised upward to 400,000. Unlike any other state in the United States, each of Nevada's 17 counties are given the legal right to decide whether or not to legalize brothel prostitution, with the exception of Clark and Washoe counties where Las Vegas and Reno (with populations greater than 400,000) are located.

Legal prostitution is kept accessible to, but out of sight of, the large metropolitan areas. The brothels are located as close to Las Vegas or Reno as is legally permitted. Several brothels are located just beyond the Clark county line in Nye County. A number of brothels are positioned along Interstate #95, which runs between Reno in the north and Las Vegas in the south (450 miles). Interstate #95 borders Nellis Air Force Base. Historically, the military has been a prolific source of johns.[30] The brothels in the north of the state are located along Interstate #80, between Reno to the west and Wells in the east (350 miles). The regular customers along this stretch of road are interstate truckers, local ranchers and miners, and businessmen from California to the west and Utah to the east.[31]

Despite the state's anti-federalist political inclinations, the United States government controls 86% of Nevada's land, more than any other U.S. state.[32] Military ranges in Nevada involve the intense and often destructive use of more than 4 million acres of Nevada's high desert land, with at least 40% of Nevada's air space designated for military use.[33] Military land use includes the Navy's bombing ranges, Air Force missile sites, National Guard tank training centers, and the Nevada Nuclear Test Site now managed by the Department of Energy.[34] For the past 50 years Nevadans have hotly debated the use of much of this land by all branches of the US military. In 2007, a blogger complained about the abuse of Nevada's land

> ...there's very little respect, throughout Nevada, for the environment and our natural resources. During my last visit, recycling was still limited, ORV [off-road vehicle] destruction unlimited, and mining cleanup almost nonexistent. ...Nevadans (in general) don't seem to care about the vast open space they've been blessed with – grazing has completely devastated most of the landscape. Anytime I was ever out and about I could only wonder what the Nevada's landscape USED to look like before all the cattle were released upon the land. Much of Nevada is cow-burnt, mowed-down, and brown...[35]

Making the connections between the land and the women, anthropologist Peggy Reeves Sanday studied interpersonal violence and environmental respect or non-respect in 156 rape-free and rape-prone cultures. Sanday found a strong relationship between the way that women were treated and the way that the land was treated. When women were treated well, so was the land, and vice versa.[36] Sanday's findings in other cultures are consistent with what seems to be the case in Nevada. Just as women in Nevada brothels are subjected to violations of their human right to safety and dignity, Nevada's land has been violated by environmental degradation. Boston's street prostitution area is called the "combat zone." Nevada's military combat zoning land use is associated with the combat zoning of prostitution/trafficking. The ecosystem destruction caused by nuclear tests and military training is paralleled by the quieter destruction of the women in the state's legal brothels.

Nevada state law gives counties almost absolute discretion on prostitution policy.[37] "Unfortunately it does not take a lot of money to buy off politicians in rural Nevada,"[38] stated a Nye County Commissioner. A pimp had offered her a $5,000 bribe. in exchange for her favorable vote on zoning his brothel. Participating in an undercover operation, the Commissioner wore a wire for the FBI, recorded the attempted bribe, and in 2006 he was indicted.[39] Historically, corruption of politicians, judges, and sheriffs has been rampant in Nevada.[40] As you will read here, the problem of corruption in Nevada remains a serious concern today.

In 2006 there were about 30 brothels in 10 Nevada counties.[41] Located in sparsely populated rural counties, the brothels reflect a lingering frontier mentality with its sex-libertarian worldview. Surrounded by barbed wire and high fences, Nevada's legal brothels are little prisons in doublewide trailers. With no resemblance to HBO's Cathouse[42] or to other pornography fantasies, the Nevada brothels are miserable institutions, some smelling worse than others, a mix of yesterday's cigarettes, sweaty clothes, semen, and last year's booze and rug cleaners.[43] Non-john observers are stunned at their first sight of the brothels. Lenore Kuo wrote,

> I cannot sufficiently underscore the intensity of the impact of the physical environment of legalized Nevada brothels. They are… shocking... [The brothels] …are surrounded by high fences, some electrified, with barbed wire or spikes on top. Entrance and egress are controlled; you must be buzzed in and out. [Two brothels] had what appeared to be an unused lookout tower, reminiscent of a prison or fort.[44]

The reality would be difficult to believe if you read it in a novel. For example, Old Bridge Brothel is owned by a Hell's Angels gang member who is nephew of a Sicilian-born pimp, himself hiding in South America to avoid arrest on tax evasion and racketeering convictions. The winding, dusty road to get there is lined with dead trees populated with live vultures. Surrounded by an 8-foot high barbed wire fence, Old Bridge is equipped with a security gate and cameras. Another legal brothel, Angel's Ladies, is owned by a mortician-turned-legal-pimp who shows visitors photos of dead people that he embalmed in his earlier career. The former mortician's wife feels sorry for johns with no money and says that she prostitutes herself to johns at no or low cost. The other women in the brothel resent the economic competition. The pimp and his exploited wife/assistant pimp/madam require the women in their brothel to say a prayer together before dinner.[45]

Regarding the legal brothels, a lobbyist said, "The less ...people think about it, the longer we are going to survive."[46] Businessmen, politicians, and the Nevada public seem to agree. Legal prostitution is recognized by many as being bad for development,[47] and is widely considered an embarrassment to the state,[48] but few are willing to confront the system. "It's permeated the minds of our city leaders; people are anesthetized to prostitution," Terri Miller explained.[49]

Poverty and economic pressure contribute to women's entry into prostitution, along with other factors. The depressed state of the US economy, a lack of educational and job opportunities for women and for people of color generally, and the increasing cultural normalization of prostitution as a "job" for poor women[50] set the stage for prostitution in Nevada. Yet the majority of women in Nevada legal prostitution want to leave it.[51]

Women in the United States earn 75% of what men do, across the board.[52] In 2005, a full-time worker at minimum wage, disproportionately female, could not afford an average priced one-bedroom apartment anywhere in the US.[53] Because of the social and legal inequality between men and women, women's earnings from prostitution are often greater than the best labor market alternatives.[54] As a committee studying poverty in Australia – where prostitution is also legal – concluded:

> Women, simply by being women, are automatically employable as prostitutes. This makes prostitution among the most accessible jobs for economically disadvantaged women who do not have a level of social security that

allows them to consistently and effectively acquire skills in order to be more widely employable.[55]

Compared to other states, women do not fare well in Nevada. The Institute for Women's Policy Research ranked women's quality of life in the 50 US states. Nevada was located 47th in proportion of women with college degrees, 40th in its employment and earnings index, 42nd in its composite health and well-being index, and worst on the list in deaths from lung cancer and from suicide.[56] In 2004, Nevada had the third-highest suicide rate in the United States.[57] That same year, Nevada was ranked the fifth-highest in the United States for suicides of 10 to 19 year-old youth.[58]

Prostitution "gives me a chance not to be one man away from homelessness again," explained a woman in a Nevada brothel.[59] "They've found out they can do better here than bagging groceries,"[60] smiled George Flint, Nevada Brothel Association lobbyist. Homeless women fare poorly in Las Vegas, which is ranked 5th on the list of the 20 "meanest" cities in the United States by the National Coalition for the Homeless.[61] Las Vegas has passed a number of ordinances targeting the homeless, including arresting those who give food to the homeless. A Catholic worker described the human tendency to help others as being "under siege" in Las Vegas, while Las Vegas Mayor Oscar Goodman called homeless advocates "enablers crying like bleating sheep."[62]

Legal prostitution is only a small sector of the sex industry in Nevada. Law enforcement sources, service providers, women in the brothels, and pimps agree that legal brothel prostitution is about 10% of all prostitution in Nevada. The other 90% is illegal — escort or outcall prostitution to hotels or casinos, in private rooms in lap dance/strip clubs, in massage parlors, or on the street. The legal brothels create a culture of prostitution in Nevada, normalizing it and attracting johns from all over the world. Las Vegas is a global sex tourism destination.

Objectification of women and commercialization of their sexuality are visibly institutionalized in the culture of Las Vegas.[63] One pimp estimated that 10,000 men arrived in Las Vegas each week seeking prostitution,[64] although prostitution is not legal in Las Vegas. The 2007 Las

Vegas Yellow Pages contain 172 pages of advertising for illegal prostitution categorized as 'entertainers' and 'massage.'

"Everyone in this town is related to the sex industry," a Las Vegas law enforcement source said. "Pit bosses, valets, bellmen, casino hosts, bartenders, taxi and limo drivers, security officers, swingers' clubs. "The demand is overwhelming," he said, explaining how one john would follow another john to the first john's hotel room and wait outside the room until she's finished with the first. I soon learned that international demand by men for women in prostitution, combined with Nevada's legalized prostitution and its illegal outgrowths make Las Vegas the epicenter of North American prostitution and trafficking - the Bangkok or Amsterdam of North America.[65]

In July 2007 a Nevada legal pimp ran his game over the San Francisco airwaves, "If people don't like legal prostitution, they can just turn a blind eye to it," he advised listeners. "There's no disease, no crime, no exploitation, and the state gets lots of taxes."[66] According to the evidence presented here, none of what the pimp said was the truth. There is overwhelming emotional and physical injury and disease caused by prostitution, lots of crime, lots of human rights violations and sexual exploitation. And there is economic exploitation of both women in prostitution and the state of Nevada. The state receives relatively low taxes compared to the cost of regulating brothels. Pimps get the big bucks, not the state and definitely not the women in prostitution.

Summary of Major Findings

Some of the major findings from our two-year study of Nevada prostitution and trafficking were:

1. **Legal prostitution's reputation far exceeds its reality**. Legal prostitution is only 10% of all prostitution in the state. Most prostitution in Nevada, including all prostitution in Las Vegas, is illegal. Prostitution is legal only in some of Nevada's rural counties which are located away from the state's two major population centers, Reno and Las Vegas.

2. **Legal prostitution does not decrease the rapes of women in Nevada. Quite the opposite.** Nevada's women are raped at rates that are twice that of New York and a fourth higher than the U.S. average. Women are three times as likely to be raped in Las Vegas as compared to New York City. Men in the state tend to normalize sexual violence. College-aged men in Nevada are much more likely than college men in other states to use women in prostitution, to go to strip clubs and massage parlors. Nevada college students tend to justify sexual exploitation and considered it acceptable that their future sons would use women in prostitution. They found it acceptable that their future daughters might become prostitutes.

3. **Prostitution and sex trafficking are linked in Nevada as elsewhere: sex trafficking happens when and where there is a demand for prostitution and a context of impunity for its customers.** In Nevada, women are trafficked primarily into the state's illegal prostitution venues: strip club prostitution, escort prostitution, and massage parlors that function as illegal brothels. There also appear to be instances where women have been trafficked into legal brothels.

4. **Despite claims to the contrary, legal prostitution does not protect women from the violence, verbal abuse, physical injury, and diseases such as HIV that occur in illegal prostitution.** Many women in the legal brothels are under intense emotional stress; many of them have symptoms of chronic institutionalization and trauma. All prostitution, even that which appears voluntary, causes harm.

5. **Legal prostitution creates a 'culture of prostitution' in the state that is fostered or tolerated by politicians, developers, and the entertainment industry.** A sex industry the size of that in Nevada exists because of political and judicial corruption and a willingness to tolerate organized crime including domestically organized motorcycle gangs and internationally organized Armenian, Russian, Israeli, Mexican, Korean, and Chinese criminals. Revenue from prostitution generated by international criminal networks has been connected with weapons trafficking.

6. **Las Vegas is the epicenter of North American prostitution and trafficking.** The sex industry in Las Vegas alone generates between one

and six billion dollars per year, according to seven informed sources. Women are trafficked for prostitution from many parts of the world into Las Vegas.

7. **There is a dangerous lack of services in Nevada for adult women seeking to escape the sex industry**, services such as emergency shelters, and social services, medical and vocational assistance. Since the prostitution of 13-17 year old children is "rampant" according to one police officer, more services for children are also needed.

8. **The links between legal and illegal prostitution in Nevada, and the profound harms caused by prostitution to all women are much like those in other countries where legal prostitution exists.** The parallels between Nevada, Australia (legal prostitution), the Netherlands (legal prostitution), and Cambodia (prostitution is illegal but socially and politically tolerated as in Las Vegas) are striking.

9. **There is a law that works. The 1999 Swedish law on prostitution has almost entirely eliminated trafficking of women into Sweden. Domestic prostitution has not increased as it has elsewhere in Europe since 2000.** Recognizing the harms of prostitution to those in it, Sweden focused on the source of the problem: men who buy women in prostitution. The victim - the woman or man or child in prostitution - is not arrested. Instead, Sweden levies felony-level charges only at buyers, pimps, and traffickers. Victims are offered services. Other countries in Europe are now adopting the Swedish law on prostitution and trafficking. The Nevada legislature is encouraged to consider this law.

2
Legal Brothel Prostitution in Nevada

Women enter legal brothel prostitution for the same reasons that women enter any other kind of prostitution: because of sex inequality, sexual abuse, poverty, lack of employment opportunities or low wages, racism, previous violent relationships, homelessness, overwhelming media messages that culturally objectify women as sex objects, and because of predators' capacity to coerce or trick women into prostitution.[67]

In Nevada, the legal brothels are zoned into rural – not urban – areas. Brothels are prohibited from operating in any urban area with a population of more than 400,000 people. Reflecting the social isolation of those in it, prostitution in Nevada is removed from the state's population centers. Whether in Turkish genelevs (walled-off multi-unit brothel complexes), Bangladeshi or Mexican prison-camp-like prostitution compounds, or Nevada brothels (ringed with barbed wire or electric fencing), women in state-zoned prostitution are physically isolated and socially rejected by the rest of society. This regulation of prostitution by zoning is a physical manifestation of the social/psychological stigma that women in prostitution feel everywhere.

Prostitution in Australia, like that in Nevada, is highly stigmatized even though it is legal. This social stigma isolates women in legal brothels, where they are generally shunned by the rest of the population, and treated disrespectfully. Their reduced social status means that the women have no one to reach out to but pimps and johns. Contacts with others are so limited that it reinforces their sense of a lack of alternatives, and the inescapability of their status, even though they are considered legal.

Women in legal prostitution fully understand society's rejection of them as complete human beings. Because of this social rejection, they sometimes hide the fact that they are prostituting. Decriminalized prostitution[68] in New Zealand was promoted as a means of removing prostitution's social stigma. But it didn't work that way there either. As one formerly prostituted woman said, "...they don't want to draw attention to themselves and what they're doing."[69] Once officially registered as prostitutes, Dutch women feared that this designation would pursue them for the rest of their lives. Despite the fact that if they officially registered as prostitutes, they would then accrue pension funds, the women still preferred anonymity.[70] Women in the Netherlands have expressed similar sentiments, even though prostitution has been legal there for many years. They specifically objected to the loss of anonymity that exists in legal prostitution since they were required to register as prostitutes. They wanted to leave prostitution as quickly as possible with no legal record of having been in prostitution.[71]

Restrictive conditions that violate human rights and dignity in the brothels

Adding to the sexual exploitation of prostitution itself, the restrictive conditions in the Nevada brothels violate human rights and dignity. A woman who prostituted in many legal Nevada brothels said that they were "…all about control. Total control. Total disrespect."[72] A witness who lived near a legal brothel told me that the women inside the brothels were "brainwashed" and "imprisoned." The women in the brothels are monitored by ubiquitous electronic surveillance. Cashiers and assistant pimps listen in on intercom systems to ensure that pimps are not cheated out of their half of the women's earnings. When women want to run an errand outside the brothel, they are usually accompanied by the pimp's escort whose time they are required to pay for.

Nevada brothel survivors and their advocates have described the legal brothels as prison-like.[73] The women inside the brothels are seen by others as a "socially untouchable class."[74] One pimp described another's brothel as "his own little prison camp" where pimps from out of state dropped off the most beaten-down, injured women for what amounted to incarceration as they were coerced under slave-like conditions to make more money by selling sex.[75] Pimps who prostituted women in cities where prostitution is illegal sometimes moved their "girls" to the legal brothels when there were police sweeps or during times when prostitution was slow business in their communities. Bluntly stating her truth, one survivor called Nevada's legal brothels, "pussy penitentiaries."

Women in brothel prostitution in other countries have taken note of the prison-like conditions of the brothel: "From the moment you're in one, you're like a prisoner." "You're like a machine."[76]

"Prison" is not only an apt metaphor for the conditions in many of Nevada's legal brothels - it is a near literal description of many of the brothels. I saw a grated iron door with bars in one brothel. The women's food was shoved through the door's steel bars between the kitchen to the brothel area.[77] "They are locked in at [name of] brothel," stated a witness who occasionally helped girls and women escape from this particular brothel. She sometimes threw food over the brothel's fence because the pimp starved the women if he thought they were too fat.[78]

Nevada brothels often have a lockdown in effect that severely limits the women's freedom to leave.[79] A woman interviewed in Douglas Lindeman's documentary Angel's Ladies explained,

> There are rules and regulations imposed on prostitutes that aren't imposed on other people. You're not allowed to have your own car here. I don't know any job in America where an employer feels that he has the right to take away your vehicle. It's like it's his own little police state.[80]

Several brothels required all the women to live on premises, and to be available for johns' use 24 hours of the day, 7 days a week. They were allowed off the premises 4 hours a week.[81]

These conditions and many other legal and illegal regulations strip women of their civil rights and their human rights. Sheriffs in some Nevada counties enforce practices that are illegal. The city of Winnemucca for example requires that prostitutes are inside brothels by 5pm and that their use of any car is "very limited." In Ely, although it is a discriminatory and illegal practice, prostitutes are not permitted in bars, and if they go to a restaurant, they must enter via a separate side entrance accompanied by a male.[82]

Pimps carefully controlled what women were permitted to say to outsiders. On an "official tour" of a brothel in southern Nevada, I asked a woman from the Czech Republic about the panic buttons. The woman became apprehensive, explaining, "we're not allowed to talk about panic buttons." Women in several of the brothels told us that when a bell was rung for a lineup for johns, they had 30 seconds to get into the line or they would be fined.

In Nevada brothel lineups, women stand in a line like cattle at an auction while johns select which he will rent.[83] Most women hate the humiliating experience of the lineups. A French survivor of prostitution described what it felt like:

> I'd freeze up inside…It was horrible, they'd look you up and down. That moment, when you felt them looking at you, sizing you up, judging you…and those men, those fat pigs who weren't worth half as much as the worst of us, they'd joke, make comments….They made you turn and face in all directions, because of course a front view wasn't enough for them. It used to make me furious, but at the same time I was panic-stricken, I didn't dare speak. I wasn't physically frightened, but it shook my confidence. I felt really [demeaned]….I was the thing he came and literally bought. He had judged me like he'd judge cattle at a fairground, and that's revolting, it's sickening, it's terrible for the women. You can't imagine it if you've never been through it yourself.[84]

In the brothels, there is pressure on women to accept any john who chooses her, regardless of how drunk, foul-smelling, verbally abusive, or threatening he seems. If a woman rejects more than one or two johns, she can be fired. One survivor described not being permitted to use her intuition regarding dangerous johns. She was allowed to reject customers only if the pimp permitted her to.[85] If women complained about conditions in the brothel, they were often blacklisted by all other brothels in the state, since there is a pimp network in Nevada with information-sharing about women who are branded troublemakers.

There is strong law enforcement support of the legal brothel system. One brothel had been owned at one point by a retired Las Vegas police officer. County sheriffs are rarely called out to the brothels because their presence would be bad for business. A brothel employee told us that on one occasion, she was afraid that a dangerously drunk john would kill another driver on the road after leaving the brothel. Refusing to let her call a taxi, he insisted on driving his car. She called the police. She was told by the brothel owners that if she ever phoned the police again for any reason she would be fired. She spoke to us under strict conditions of anonymity, fearing for her safety.

Women in the legal brothels don't earn the mythical $15,000 a week that pimps brag about on talk shows. Women are financially exploited in the legal brothels just like women in illegal sectors of the sex industry. The legal brothels take 50% of their earnings.[86] In addition they are required to tip brothel employees (cooks, assistant pimps, bartenders, bouncers), pay rent for rooms in which they both live and turn tricks, buy food, pay for medical check-ups and all medical care. Since the women are considered "private contractors" by the state, they receive no employment benefits, and no health insurance. After the pimp's 50% cut is taken, and after tips and fees and food and lodging are paid for, her share of the earning from prostitution is about one fifth of the money she brings in.[87] Brothel owner-pimps charged inflated prices for everything inside the brothel: rent, tips to every employee, and most of all, food.[88] Louise Brown describes this same process of gouging women on prices for food and other essentials by brothel owners in Cambodia.[89]

Women in the brothels were gouged on rent. One brothel charged $900 a month for a tiny room in a trailer in the desert. Women's prostitution earnings are held by the pimp. If she leaves prior to an agreed-upon 2 or 3 week period of time, $1500 - 2500 is deducted from her pay as a fine. This amount is large enough to function as economic coercion and it keeps most women in the brothel even though they may prefer to leave. Pimps' economic control sometimes includes preventing her from access to the ownership papers and keys to her car. These are the same techniques used by battering husbands to control their wives.[90]

Larger breasts are associated with higher incomes in the sex industry. One woman sardonically described being in "debt bondage via breast enlargement surgery" to her pimp, because he gave her an advance for the surgery. Age on the other hand, increases prostituted women's economic vulnerability. A 60 year-old woman who prostituted at a legal brothel was seen running into the parking lot after johns, offering lower prices.[91]

One of the myths about legal prostitution is that it makes prostitution safer. There is no evidence supporting this myth. In fact, there is much evidence that legal prostitution, like illegal prostitution, is physically dangerous. Sociologists Barbara Brents and Kate Hausbeck asked women if they felt safe in legal prostitution and many responded affirmatively. Usually, however, by "safe," women mean safe in comparison to other prostitution. The concept of safety is relative, given that all prostitution is associated with a high likelihood of violence.[92]

In a study of prostitution in 9 countries, we found that many respondents did not feel that they would be safer from physical and sexual assault if prostitution were legal. Forty six percent of people in prostitution in 6 countries felt that they were no safer from physical and sexual assault if prostitution were legal. Half of the 100 prostituting respondents in a separate study in Washington, D.C. expressed the same views.[93] In an indictment of legal prostitution, more than half of German women in legal prostitution told us that they would be no safer in legal as compared to illegal prostitution.[94]

Four of Australia's six states and territories, like 10 of Nevada's counties, have legal prostitution. The Australian Occupational and Safety Guidelines recommend self defense classes for women in prostitution, and they also promote classes in hostage negotiation skills. The implication is that if you are prostituting, you stand a good chance of being kidnapped, therefore you need to learn how to artfully negotiate with the john who kidnaps you.

Sometimes lethal violence occurs in legal brothels. A woman interviewed by Brents and Hausbeck described a near-lethal assault by a john in a brothel where he cornered and choked her, fracturing her larynx. She stated that she would probably be dead if another woman hadn't heard the scuffle and broken into her room.[95] Another woman they interviewed spoke about the need to "keep a weapon an arm's length away" in the brothel.[96] A brothel owner in the Netherlands complained about a well-intentioned but naive ordinance requiring Dutch brothels to have pillows in their rooms. He commented, "You don't want a pillow in the [brothel's] room. It's a murder weapon." Familiar with the hostility that johns directed towards women in prostitution, this Dutch pimp understood that johns are regularly murderous toward women.[97]

Most of the Nevada brothels had panic buttons in the women's rooms. Yet physical or sexual violence occurs in the blink of an eye, before anyone can save her. Although the women in the Brents and Hausbeck study reported that they *felt* safer in a legal brothel than they did in illegal prostitution, panic buttons in the room served more as a "symbolic than an actual mechanism for protection against danger."[98] In 2004, a woman prostituting at a Nevada brothel filed civil lawsuits against a john who assaulted her and also against the brothel's pimp who failed to call police. She stated that the panic button in her room was not working.[99]

Legal prostitution exploits women's lack of alternatives for economic survival. A woman in a Nevada brothel explained to a delegation visiting from India, "If you have any small option at all, you leave here."[100] Economic stress, race and sex discrimination, and emotional and sexual abuse all channel women into prostitution. We don't really know what "choice" means until women can freely choose from an array of educational and vocational options. Discussing the ferocious economic pressure on women that contributes to their entry into prostitution, one woman described arriving in the United States from the Dominican Republic with no English. Attempting to support her 5 year old son, she was not able to pay her rent with the income she received from Wendy's. She moved to Nevada to prostitute in a legal brothel.

We know that battered women, prisoners of war, and women in legal prostitution, minimize the harm and try to look on the positive side as a survival mechanism. Witnesses may minimize the harm of prostitution because it's too painful to hear about the truth of women's experiences. Once in prostitution, a woman explained the need to justify that.

> There's a protective denial. You have to convince yourself and everyone around you that it's great. You tell the lie – "I like it" – so much that you believe it yourself. You make it OK by saying, 'I haven't been beat up today. I haven't lost all my money today.' Women have to justify it: they can't tell themselves or anyone else the reality of it or else they'd die. In the brothels, there's the illusion of safety. Then there's the downward spiral to… street. Until you step out of prostitution, you don't realize what the emotional wreckage is.[101]

Interviews with 45 women in Nevada's legal brothels

I wanted to hear from the women themselves what their experiences of Nevada legal brothel prostitution were. I knew that they would minimize how bad it was, not only to make prostitution seem like a reasonable job choice to the interviewers, but especially to justify it to themselves. The women in the brothels reduced the cognitive dissonance of permitting strangers to sexually abuse them for pay by doing what we all do every day: they ignored bad things or they pretended that unpleasantness will soon go away, or they called the degrading abuse of prostitution by another name that sounded better.[102] The rationale was that money trumped all.

We asked women for the often painful truths about their experiences. We let them know that if they wanted to terminate the interview, or not answer questions – that was their right and it was OK with us. Almost everyone wanted to complete the interview. We were asking the women to briefly remove a mask that was crucial to their psychological survival. In prostitution, women act the part that the johns want them to play. The deception is calculated and deliberate. One woman watched football games and stock car races in the back room of a brothel so she could learn how to talk to johns about their favorite topics. The johns frequently believe that her act, her prostitute persona, is who she really is. Yet for women in prostitution to be truly known by johns and pimps is to be further violated and further exploited. All she has of her real self is her real name which she carefully guards from johns. She also hides those private facts about who she is such as where she lives, where she likes to go for a walk, her kids' names and those things that make up her personhood.

Researchers have generally had a difficult time interviewing women in indoor prostitution. Pimps often prohibit independent researchers' entry. For example, Tooru Nemoto and colleagues[103] interviewed women in San Francisco massage brothels and found that many had been verbally, physically or sexually abused. Yet his data was based only on interviews with half of the women that he attempted to interview in San Francisco's brothels. As was the case in our study of legal brothels in Nevada, half of the illegal brothels in San Francisco would not permit Nemoto's research team to interview women in their brothels.

We[104] interviewed 45 women in 8 Nevada brothels - 3 brothels in southern Nevada and 5 in northern Nevada. We visited an additional 6 brothels but we were not permitted to speak with any of the women inside. The women spoke to us under challenging, often adversarial conditions. The pimps didn't want the women telling us anything negative about "the business." Given the circumstances of the interviews, most of our data offer a conservative perspective on the dangers of legal prostitution.

At those brothels where we were not permitted to speak with the women, we were sometimes allowed to speak with owners or assistant pimps/madams. We interviewed them, too. One of the brothels that refused to allow us to speak with the women, physically separated the brothel bedrooms from the brothel restaurant using a grated iron door that looked just like a jail cell. The women were locked inside this particular brothel. Accounts from neighbors and other witnesses corroborated the captivity of women in this particular brothel.[105]

A layer of female control of women in prostitution ensures the day-to-day functioning of the brothel. Most of the legal brothels were managed by women assistant pimps or madams who had sometimes aged out of prostitution or who could not find other jobs in the rural economy. These women enforced the pimp's rules with a vengeance. Their role in the Nevada legal brothels is much the same as that of the Mexican women who facilitate trafficking from Mexico to the United States. Landesman described women's role in trafficking young women from Mexico to the United States. Once seduced by men or boys, the women are then turned over to "the mothers" who indoctrinate them, "handling" them, exerting physical violence and psychological torture. Because women can more easily gain young girls' trust, they can more easily crush them.[106]

The brothels are not named here out of concern for the physical safety of the women who are quoted. Because of the prison-like nature of Nevada's legal brothels, and because they are located in rural areas without reasonable access to medical emergency services, alternative housing, or rape crisis centers, we did not want to worsen the conditions under which many of these women already lived.

Aware of listening devices in all rooms of the brothels, many women spoke to us in whispers. They knew that pimps would be unhappy if they provided information that reflected badly on the legal brothel system in Nevada. A woman I interviewed turned a TV to high volume but still whispered while she spoke about her trafficking and her fears about cervical cancer. After 30 minutes of a whispered discussion with this woman, my research associate suddenly entered the room, loudly commenting about the weather. She later told me that she had seen an assistant pimp hiding behind a potted plant trying to listen in on the interview. She interrupted the interview until the assistant pimp backed off.

A woman who had prostituted in a Nevada brothel told us that the experience was like being the pimp's property for two weeks. "You have sex when they want, with whom they want, and it doesn't matter how you feel or anything. You're locked in a box for two weeks and guys come in and out."

Several other researchers have investigated the conditions women are subject to in Nevada brothels. Psychiatrist Alexa Albert actually lived in a Nevada brothel for an extended period of time, getting to know a number of the women quite well. Her book *Brothel* [107] was an in-depth look at the women's lives, including their submission to pimps, sexual exploitation, and psychological stress. Some of our findings were consistent with Albert's observations. Sociologists Barbara Brents and Kate Hausbeck surveyed 40 women in the Nevada brothels between 1998 and 2002. They have written about the history of legal prostitution in Nevada and have described violence in the brothels. Lenore Kuo investigated legal prostitution in 3 Nevada brothels in the mid-1990's, interviewing several women in prostitution as well as sheriffs and pimps.

Most US agencies offering services to women escaping prostitution have worked with women from the Nevada brothels. When women escape from the brothels, they flee to other states: California, Oregon, Arizona, among others. If they are lucky, they find the Council for Prostitution Alternatives in Portland, Sex Workers Anonymous in Las Vegas, Dignity House in Phoenix, Breaking Free in Minneapolis, Mary Magdalene Project in southern California. We interviewed staff and clients from these agencies.

I spoke with a 25-year old woman I'll call Amber in a Nevada brothel who had grown up in a sexually and emotionally abusive family. She had been diagnosed in childhood with a learning disorder. Implying that Amber was defective because she couldn't read or write, her mother sold 16 year-old Amber into a legal brothel. Her parents then received child welfare payments from social services for her three children. Amber felt that her role in life was to turn tricks. "But they told me I'd be a bad parent," Amber told me when I brought up the possibility of her leaving the brothel and raising the children herself, with social service payments going to her rather than her parents. The depth of her self-loathing was off the charts. She was deeply convinced that the only way she could do anything for her children, aged 8, 5, and 2 years, was to prostitute and give that money to her parents. Not surprisingly, Amber had been in and out of mental hospitals several times, and when I saw her she was on 9 psychotropic medications including sleeping pills, antidepressants, antianxiety medications, antipsychotics, and drugs to reduce the severe side effects of antipsychotics. Given all the drugs she was on, I was surprised that she could walk and talk. Despite the drugs, she still had symptoms of posttraumatic stress disorder.

Amber also suffered from abscessed breasts as a result of johns' dirty mouths. She told me about a nightmare in which she was in a motel with a chain around her ankle. When she woke up, she thought, damn, I'm going back to the brothel.[108]

The 45 women we interviewed were diverse in terms of ethnicity, culture, and nationality. Forty nine percent described themselves as white European-American, 15% Latina (including Mexican, Mexican-American, Dominican Republic, Caribbean, Brazilian, French-Spanish), 13% were African American, 9% Native American, 7% Multiracial/Multicultural, 2% Cambodian, 2% Chinese, 2% Filipina. These percentages would have changed dramatically if we had included the approximately 30 Chinese women we located in one brothel, but we were not permitted to interview any of them so they are not included here.

Johns select women in prostitution in part based on the women's skin color and ethnicity, expressing their preferences for racially stereotyped women: passive Asian servants, wild Latinas, animal-like Africans, naïve Irish lassies. However, their awareness of the women's real-world ethnicity is often confused and racist. Any Asian woman seems to be called a "geisha girl" even though the term refers specifically to Japanese women in prostitution. A Vietnamese-American-owned brothel in Nevada advertised as Geisha Girl is owned by a pimp who manages illegal massage parlors and nail salons in California. A Korean-American woman was sold as a "geisha girl" and was pressured by pimps to dress in Japanese kimonos. Although she was raised in the United States and spoke perfect English, she was coerced by pimps into speaking broken English, because johns eroticized their fantasy of the vulnerable whore who has no avenue of escape and who will be forever grateful for his "rescue" (that is, his payment for her prostitution) of her. Another woman wrote that she deliberately tried to sound painfully stupid in the presence of johns, since they needed to know they would not be outsmarted.[109]

The sex industry is racist with respect to African American women. Mainstream pornography only features light-skinned African American women, and then only rarely, and that is what johns seek if they buy an African American woman at all. "Black adult performers have to be twice as polished and twice as ingratiating as their white counterparts, and even then, they make far less money...."[110]

The women we interviewed in the brothels averaged 29 years of age. Their ages ranged from 18 to 56 years. See Table 1. They had been in prostitution for an average of more than seven years before we interviewed them.

Table 1. Demographics of 45 Women in Nevada Legal Prostitution

Mean Age	29.7
Age range	18-56
Mean age of entry into prostitution	22.4
Mean number of years in prostitution	7.5
Was prostituted as a child (under18)	23%
Total number of johns serviced in prostitution	26-200 johns 24% 201-600 johns 21% 601-1000 johns 19% Over 1000 johns 36%
States she wants to escape prostitution	81%

Many adult women who are in prostitution started out as children. As poverty increases and educational opportunities and public supports for children and their mothers decrease, experts agree that girls are entering prostitution at younger and younger ages. Across the United States the average age that girls enter prostitution is 13-14 years. This was also the age when many women in the legal brothels started prostituting. Nearly one-fourth of our interviewees reported being prostituted as children. The prostitution of children is child sexual assault according to the law, but it is generally not thought of that way. Adolescent girls who prostitute are considered delinquent or bad rather than as victims of sexual exploitation and abuse. Even the women themselves probably did not consider their prostitution as children to be in the category of childhood sexual abuse.

The single most difficult question for the women to answer was one in which they were asked to check off the total number of customers they'd ever seen in prostitution. With 12 options, the numbers of johns went from "1-25" all the way up to "more than 1000." This question was one of Andrea Dworkin's many contributions to our questionnaire. "More than any other item," she told me, "this one number will give you an index of the harm suffered by women in prostitution." The women felt this. Several shuddered when they saw the question. One woman said, "I don't want to add it up. If I thought about it as much as you, I wouldn't be doing it. But then, I'd be in a car starving…" Others simply refused to answer, with one woman underlining her response which carried a double meaning, "I <u>do not count</u>." A woman in another country (not part of this research) who'd prostituted in a Moroccan legal brothel explained that pimps controlled and disciplined the women by increasing the pace at which they sent her johns to service in prostitution.[111]

The longer women are in prostitution, the more physical danger they are exposed to, and the more they suffer emotional and physical harm.[112] These 45 women had serviced many hundreds of men in their lifetimes. Only 24% had been prostituted to fewer than 200 johns, with 76% of our interviewees prostituted to more than 200 men. 36% of the women we interviewed had been prostituted by more than 1000 men. I am confident that the actual numbers of johns using these women was much higher, since it was extremely painful for the women to acknowledge just how many men had used them.

In 9 countries on 5 continents, 89% of more than 850 women in prostitution told us that they wanted to get out.[113]   Most of the women we interviewed in Nevada's brothels told us the same thing. 81% of the Nevada women told us that they wanted to escape prostitution, regardless of its legal status. "It's all the same emotionally, no matter where we work," said one woman, referring to other – illegal - locations where she had been rented for sex. In addition to prostituting in legal brothels, half of these 45 women had also prostituted in strip club/lapdance clubs, another half in escort prostitution. Many had prostituted in illegal massage parlors, street prostitution, phone sex, table dance clubs, peep shows, and military bases. See Table 2.

Table 2. Types of Non-brothel Prostitution Engaged in by 45 Women in Nevada Legal Brothels

Strip club, lapdancing	51%
Escort	49%
Pornography	47%
Massage	27%
Street	24%
Phone sex	20%
Table dance club	16%
Internet	16%
Other types of prostitution[114]	16%
Peep show	7%

Some women were afraid that if they reported violence from johns, they themselves might be blamed for it or even fired. This was confirmed by a lengthy interview with a former brothel manager who told me that she feared for her life if her identity was revealed. She stated that only a small percentage of brothel violence is reported and that the women are so accustomed to violence in their lives that an assault from a john seemed "almost insignificant" to them. I was reminded of one woman's explanation of what prostitution was like, "what is rape to others, is normal for us."[115]

In fact violence against all women in Nevada is grossly underreported. Public health agencies in Nevada[116] keep poor or no records of many of the kinds of interpersonal violence experienced by women in prostitution. As a result, researching the domestic violence in prostitution and trafficking becomes more challenging. For example, as of 2007, Nevada did not require emergency rooms to report non-fatal injury data except for firearms-related injuries. The many women who have been beaten, raped, or strangled by buyers or pimps - and whose injuries were actually documented by ERs - are not noted in violence databases. Nevada does not participate in the DAWN network, which documents substance abuse overdoses including those related to attempted suicide. This type of record is especially important since a significant number of prostituted and trafficked women attempt or commit suicide. The practice of psychological autopsy, which would also be helpful in determining how many women have killed themselves because of their inability to escape prostitution or trafficking, is not standard practice in Nevada, although it should be.[117]

Almost half of the women we interviewed in the brothels had been homeless at some point in their lives. Their lack of stable housing reflects their lack of a real alternatives and the economic hardship that propels women into prostitution.[118] Another myth about legal prostitution is that it protects women from pimps, exploitation, and violence. Yet more than a fourth of our interviewees told us that they had been coerced or pressured into performing an act of prostitution that they did not choose to perform. 15% had been threatened with weapons, and 24% had been physically assaulted.

Table 3. Violence in the Lives of 45 Women in Legal Nevada Prostitution

Gave part or all of income to pimp/partner	57% (25 of 44)
Report that women in brothels have pimps	50% (22 of 44)
Had pornography made of her in prostitution	47% (21 of 45)
Current or past homelessness	47% (21 of 45)
Verbally abused in prostitution	44% (19 of 45)
Coerced or pressured into act of prostitution	27% (12 of 44)
Physically assaulted in prostitution	24% (10 of 42)
Upset by an attempt to make her do what was seen in pornography	20% (9 of 45)
Threatened with a weapon in prostitution	15% (6 of 41)

Historically, women in the Nevada brothels were actually required by the legal brothel pimps to also have pimps outside the brothel. Today, most of the women in the brothels are still pimped in, often by pimps called boyfriends, husbands, or friends.[119]   Researchers and advocates have observed that almost all of the women in the legal brothels have pimps.[120] This contradicts a widespread misconception that prostitution in the Nevada legal brothels is entirely voluntary and that the women inside the brothels freely choose to go there.

Although reluctant to admit that their boyfriends and husbands were pimping them, half of the women we spoke to acknowledged that they knew others who had illegal pimps outside the legal brothels. They told us that at least 50% of the women in the brothels were under the control of illegal pimps outside the brothels, and 57% were giving all or part of their earnings to someone other than the legal brothel's pimp. See Table 3. Like women pimped in illegal prostitution, our interviewees had ever-increasing earning quotas set by pimps. Some had their pimps' brands tattooed on their bodies.[121]

The numbers here are conservative, and the amount of coercion and violence is probably higher than this figure suggests. What is remarkable is that even with pimps roaming the premises and listening devices in interview rooms, the women told us about as much violence in their lives as they did. Another 20% of the women had been pressured or coerced by johns into performing an act he wanted her to imitate based on his viewing of pornography. See Table 3.

Symptoms of Psychological Distress of Women in Legal Brothels

The same behaviors used by batterers against their domestic partners are used by pimps to control women in Nevada brothels: social isolation, control of their cash and assets, verbal abuse, sexual and physical violence.[122] Brothel owners sometimes call in the women's outside pimps to enforce discipline, and these men frequently use brute force. The Nevada state-sponsored brothel-owning pimps benefit from illegal pimps' "relentless demands that women earn more and more money."[123] Several women told us that their pimps placed them in the Nevada brothels as punishment if they were not earning enough elsewhere. The legal brothels owned by the most abusive pimps were well known to both the women and their outside pimps for operating their legal brothels with absolute, often violent, control. One brothel owner starved women if he decided that they were too fat.

Traumatic bonding keeps people in relationships with abusive partners. Just as women seemingly inexplicably return to their batterers, and abducted children fail to run away from sexual abuse perpetrators - prostituted women's return to violent abusers must be understood as a way of staying alive under psychological conditions of physical and emotional captivity and domestic terrorism. A survivor of both legal and illegal Nevada prostitution described the brutal psychological process of being coerced into "recant to pimps or masters." In this practice, girls and young women are deliberately terrorized by being forced to witness someone else being beaten and even murdered. This woman told us that after prolonged witnessing of others' torture and abuse, she was eventually programmed to believe that that only way she would avoid torture is to become the pimp/master's puppet.

Pimps take advantage of the fact that women who have long histories of violent abuse, family neglect, sexual assaults from partners and husbands, and who suffer the emotional and mental consequences of these abuses are particularly vulnerable to entering, and being kept in, prostitution. "Most women [in his legal brothel] have been sexually abused as kids, some are bipolar, some are schizophrenic," a pimp explained to us matter-of-factly. We heard the same assessment from advocates for the women. These serious mental disorders are both cause and effect. Mental distress or disability are factors which leave a person vulnerable to sexual predation and these are later exacerbated by prostitution itself.

A surprisingly low percentage - 33% - of our interviewees reported sexual abuse in childhood. This percentage is lower than the likely actual incidence of sexual abuse because of the symptoms of numbing, avoidance, and dissociation among these women and also because of the discomfort of talking about personal history with listening devices or pimps attempting to overhear. The actual rate of childhood sexual abuse among women in the Nevada legal brothels is likely to be significantly higher than 33%. Other researchers and advocates have documented rates of childhood sexual abuse of 70% to 85% among women in prostitution.[124]

The verbal abuse that is commonplace in prostitution is a source of serious and long-lasting harm even though verbal sexual abuse is assumed to be part of the job description of prostitution. Eighty-eight percent of women in a recent study described verbal abuse as being intrinsic to prostitution.[125]

In addition to racist and sexist verbal abuse from johns, sophisticated psychological coercion and control is used by pimps and traffickers.[126] Seasoning for prostitution is aimed at convincing a woman that her primary value is as a sex commodity. When this indoctrination process is complete, women appear to embrace their identity as sex workers. This is as true for women in the legal brothels as it is for women controlled by the most violent street pimp. While in the brothels, women go to great lengths to pacify themselves when they can not access other alternatives to economic survival. In one brothel, a woman told us, "Because no one really enjoys getting sold, you keep in mind: I'm getting paid, I'm getting paid, I'm getting paid."

Speaking about how it felt to prostitute in a Nevada brothel, a woman interviewed by Albert said, "The first words that come to mind are: degraded, dehumanized, used, victim, ashamed, humiliated, embarrassed, insulted, slave, rape, violated." [127]

"It's like you sign a contract to be raped," explained a woman we interviewed.

"The first six months, I cried all the time," said another woman.

"It makes you feel like you're not worth loving or marrying," explained a third woman.

The women in the brothels spoke about their intense need to disappear psychologically in order to survive these painful emotions. "You can't be mentally present in this business or you'll go crazy," another woman told us.

Dissociation is a mental tuning-out to avoid unbearable and inescapable stress. A dissociative response to prostitution is an extreme version of the denial that happens every day in most peoples' lives: bad things are ignored, or we pretend they will go away, or we call them by another name.[128] Incest victims, rape victims, and prisoners-of-war dissociate in order to mentally survive. The same dissociation that is used to survive incest is later used to survive prostitution. Incest is boot camp for prostitution.[129] The dissociation manages the overwhelming fear and pain of prostitution. It mitigates the john's cruelty by splitting that experience off from the rest of her self.

Dissociation is studied and learned in prostitution. One woman said, "It's like putting up a block in your mind. You go through the motions, but you're not really there, you're taking a trip out of the room."[130] "In order to do this, you need to not be thinking," another woman explained.

Over time, the objectification by pimps and johns are internalized by women in prostitution. Parts of her body are numbed and compartmentalized. Starting to think about herself the same way that johns and pimps do, she eventually also views her body as a commodity, rather than as integral to the rest of her self. Other trauma and torture survivors experience this same disconnectedness.[131]

Most women say that they cannot prostitute unless they dissociate.. A woman in Australian prostitution felt much like the women in the Nevada brothels:

> When I work my face is expressionless. Of course when I first speak to them I put on my best pretty-please smile. *But in the room I become nothing and nobody.* I can only work from below the neck. If I have to think about a service or involve my mind even slightly, I feel dirty. I avoid fantasies. I don't want to participate in their filth.[132]

The women we interviewed reported high levels of dissociation. This strongly suggests that they were experiencing chronic, unrelenting, and overwhelming stress in the brothels. A woman told us, "When I'm alone I go into a trance." Another said, "I was in a brothel, laying down, sleeping and I said 'O.K., you're in bed, sleeping. You're not dead.'"

A woman interviewed in a Nevada brothel by another researcher described disconnecting, feeling nothing, and seeing "nothingness where the man's face should be."[133] Las Vegas psychologist Howard Meadow observed that the women in the brothels "know how to flip the switch off."[134] Christine Stark told us that the women in the Nevada brothels were completely isolated from any real human contact, that they lived, cage-like, outside normal society. She observed that as a result of these split-off lives in the rural brothels, the women later had great difficulty with normal social interactions.[135]

The Thomas Cook Travel Agency advertised what they described as a "zoo-like" experience of driving through Amsterdam's legal prostitution areas where women are sold out of windows to tourist johns. One can barely imagine the dehumanizing contempt that is aimed at the women who are stared at by tourists like animals in a zoo.[136] A staff member from an Oregon agency offering alternatives to prostitution observed that women leaving the Nevada legal brothels were different from other prostitution survivors. Describing the women in legal brothels as more passive and more institutionalized, she felt it was a result of their having lived in the brothels like prisoners or animals in a zoo, with fresh-air breaks permitted only in enclosed, fenced-off areas.[137]

The women's institutionalization can also be understood as a state of moderate and chronic dissociation. Accommodation to conditions of living inside a brothel happens in Nevada, just as it happens in India. The brothel becomes home, a safe place, even one's entire world if she is there long enough.[138] A study of the mental and physical symptoms of trafficking victims in Europe also found that psychological symptoms of numbing, tuning out, and dissociation were extremely common. Memory problems including forgetting the past and difficulty with parts of the past were among the most severe and long lasting symptoms.[139]

Most have underestimated the dissociation required to prostitute for extended periods of time. Dissociation may be incorrectly assumed to be manipulative or deliberately deceitful.[140] Some people are confused by dissociation, failing to grasp that the dissociation they see as acceptance of prostitution or streetwise bravado or seductiveness is actually a consequence of multiple traumatic factors, including childhood abuse, childhood mental conditioning, and in prostitution: an additional layer of traumatic stress and mind control by pimps.

Dissociation, emotional numbing, and prostitution were statistically associated among the women in the brothels. Our analyses showed that the more johns a woman had serviced, the more she also reported that she had memory problems in recalling details of that past.[141] And the longer women had been in prostitution, the more they avoided situations that reminded them of the traumatic past.[142] After many years in prostitution, the women in the brothels shut down so completely that they felt emotionally numb around people with whom they had previously been close.[143]

Despite the women's dissociation and despite their attempts to avoid traumatic memories of the past and despite their emotional numbing, they were still unable to avoid intensely disturbing flashbacks of past traumatic events. The longer they had been in prostitution, the more likely they were to experience flashbacks.[144]

Almost half of the women we interviewed had pornography made of them while in prostitution despite the fact that being filmed in prostitution for the purpose of producing pornography was a source of the women's increased distress. One pimp specialized in hiring women who had made pornography videos. He then used their pornography videos to promote his brothel, advertising "pornstars for rent." Women new to prostitution reported intense pressure from this particular pimp to be filmed by him, even though they often preferred not to have that documentation made of their prostitution.[145] At this particular brothel, women told us it was not possible to make much money unless they permitted the pimp to film their prostitution. Once their prostitution was documented via pornography, women felt it both defined and shamed them. In the long run, the pornography made it more difficult to escape the sex industry since it was a permanent record of their prostitution circulating on the Internet or in adult stores.

Johns frequently seek to buy specific acts of prostitution by pointing out what they want in pornography. "See this? Do that." Many of the women we interviewed felt pressured by pimps to perform acts for johns that the johns saw in pornography. Although the women smile and act as if they were enjoying the "sex," they in fact experienced symptoms of traumatic stress related to pornography production and its use by johns. Women in the legal brothels whose johns demanded that they imitate pornography experienced significantly more traumatic flashbacks than women whose johns did not require imitation of pornography.[146]

Compared to those women who did not have pornography made of their prostitution, the women in the legal brothels who had pornography made of them reported significantly higher levels of emotional distress when events in their lives triggered reminders of past trauma.[147] Most prostitution transactions during which a woman is treated as an object, during which her feelings are ignored, and during which she is sexually exploited/abused – trigger memories of past sexual and emotional trauma. In prostitution, the emotional wounds of her past are constantly being reenacted and exacerbated rather than healed.

Sometimes dissociation results in enthusiastic endorsement of prostitution as a sexy, lucrative job. Playwright Carolyn Gage has written about the relation between incest, dissociation, and advocacy of prostitution in the life of one woman.

> Angie had sexually serviced, she estimated, about two thousand men. She owned a home, which she referred to as 'the house that fucking built.' As a prostitute, Angie had become a spokeswoman for prostitution. She described herself as a 'poster child' for liberal organizations advocating for legalization of prostitution. She was, apparently, their model of the happy, healthy hooker.

> Angie's prostitution was socially supported and paid well. To understand herself as a former child victim would be to see that her seemingly autonomous, even rebellious choices were, in fact, programmed responses to previous torture and captivity. The elements of choice and free will so critical to her sense of personhood were not as she had seen them. With every act of so-called sexual liberation, she was reinscribing her trauma.

> For three decades, Angie had had no memories of her sexual abuse as a child. Growing up in the Midwest as the only child of Christian fundamentalist parents, she had not remembered anything extraordinary about her childhood...Later, she married and began to participate in group sex and partner-swapping. It was the Sixties, and Angie considered herself liberated....

> The dissociated identity has a profound investment in denying that it is split off, because the original stakes were usually nothing less than survival. For this reason, the dissociated personality can be very persuasive. When Angie said she loved being a prostitute, loved servicing her clients, would have done it even without pay, she was persuasive

because she believed it – and because she believed it, she was very credible... (Carolyn Gage, in press)

Angie's memories of chronic sexual abuse returned only after she had stopped prostitution. Until that time, the memories of childhood abuse were completely split off from her normal consciousness. Later, she met a supportive friend, and took a class in which she began to write about her life. At this point, memories of the sexual abuse surfaced. For a time, she felt that she had betrayed other women by her previous advocacy of prostitution as a glamorous and lucrative career choice.

Nevada Legal Prostitution and HIV

In legal prostitution, there is an intention on the part of the state to control the threat of HIV to the community of wives and girlfriends. The threat is created through men's use of women in prostitution. In Australia, HIV harm reduction strategy has not been successful because legalized prostitution failed to shift the power imbalance between men and the women they bought in prostitution, and it also failed to decrease men's demand for unsafe sex.[148]

There is no empirical evidence that legal prostitution decreases HIV.[149] Yet it is widely assumed that Nevada brothel prostitution is "100% safe." Beginning in 1987 prostitutes in Nevada brothels were required by state law to be tested for sexually transmitted diseases every week and tested for HIV once a month. Nevada state law enshrines the notion that "if *clean* women are offered up to johns, then hey, what's all the fuss about prostitution?" The purpose of the following discussion is not to blame women in prostitution but to show that once in prostitution, it is almost impossible to protect her from HIV or from other violence from johns.

Women are stigmatized as 'vectors of disease' when she is required to be tested while the john remains untested. Johns on the other hand – the economic driving forces of the sex industry – remain socially and legally invisible. The medieval practice of HIV testing of women but not men is a hair-breadth away from reactionary sexist visions of women's genitals as filthy or evil and women themselves as inferior to men. This sexually retrogressive view of women is like that of the pornography produced in the brothels which similarly defines women as stupid sluts who enjoy being humiliated by men.

A popular myth is that Nevada has never had a case of HIV transmission linked to a brothel. Brothel owners in Nevada, like pimps everywhere else, advertise "clean girls" who are HIV-and-disease-free. This information is promoted by many people in the state. For example,

> Since HIV testing began in 1986, there has not been one positive test among the state's prostitutes, according to Randall Todd, chief of the Nevada State Health Division's Bureau of Disease Control and Intervention Services.

"Logic dictates," said Nevada brothel prostitution survivor Laura Anderson, that if one is genuinely concerned with public health, it is more important to administer STD tests to customers than to prostitutes."[150] Testing women but not their customers for HIV fosters the medical prejudice that the woman and not the john is the source of infection. The evidence suggests that the opposite is the case. HIV is transmitted from john to prostitute via vaginal and anal intercourse.[151] Johns' rapes of prostituted women and their refusal to use condoms are the primary sources of HIV infection among the women.

Current science regarding HIV is that women with multiple partners are at highest risk.[152] The greater the number of sex partners, the higher that person's risk for HIV.[153] Since the women in Nevada brothels have many sex partners, some having serviced thousands of johns, they are at extremely high risk for HIV.

Women in the legal brothels felt that they were considered expendable resources. Blunt and on point, a woman from a legal brothel said, "Why save the 'ho? We have to be pure, clean and healthy but any man can walk in here and we know nothing about his health history."[154] The HIV testing requirement is thinly disguised advocacy for clean meat for johns.

Nevada law shows little interest in the health of the women in its brothels. The medical testing required of prostituted women working in brothels is aimed at promoting the business of prostitution rather than at protecting women in prostitution. A law in Tijuana Mexico is based on this same reasoning. The Tijuana law requires that HIV test results be printed on health certificates of women in prostitution. The law was promoted by the city's 236 owners of strip clubs and tourist bars, whose "principal interest is that our clients are not infected."[155]

If a woman in a Nevada brothel tests positive for HIV, she is fired. She may become involved in illegal prostitution either in bars or hotels in the small towns or in Las Vegas, where prostitution is illegal. If she has HIV and if she is arrested for solicitation of prostitution, she can be charged under state law with attempted murder. Yet Nevada state law does not similarly penalize the john who is HIV positive. Johns are not tested for HIV, HPV, or hepatitis - all diseases of great concern to women in prostitution. Asked about the lack of testing of johns, a pimp stated that at his brothel there were "too many johns to test," but there were fewer women who were more "controlled" who could be tested.[156] Johns are not "controllable" by pimps or by the state, and least of all by women in prostitution.

Legal prostitution advocates emphasize the frequent and regular *testing* for HIV that occurs in the legal brothels, but fail to mention that it takes three months for the test to confirm whether or not HIV is present.[157] At any point after an HIV test, a woman can be infected by a john but the test would not show positive results for three months. Lenore Kuo noted that "It is quite possible that a prostitute can be exposed to an STD [including HIV] since her most recent test, particularly since three months must pass between the time of exposure and detection by currently employed tests for HIV."[158]

In Nevada's legal brothels the actual rate of HIV is unclear, as is the case in Tijuana where a similar testing program is in effect. The mayor of Tijuana, a businessman, stated that only 3 cases of HIV had ever been detected since the city began testing 5000 women in prostitution a month. Thus the mayor is suggesting an HIV prevalence of close to 0% among those in Tijuana prostitution. Disagreeing with the mayor's claims which in effect amount to a handshake with pimps, a Tijuana physician who ran a state-sponsored health clinic offering free HIV testing estimated that 10-12% of those in the Tijuana sex trade tested positive for HIV.[159]

There is evidence of HIV infection among women in Nevada prostitution. According to Dr. Karen Gedney, Senior Physician of the Nevada Department of Corrections: since 1985 Nevada's state correctional system screens all women and men entering the state's prisons for HIV. Twice as many women as men have HIV in the Nevada prison system. In 1995, 5% of all women entering the Nevada prisons were infected with HIV, with a majority having histories of prostitution arrests.[160] Las Vegas Metro police also report a 5% rate of HIV among women arrested for prostitution.[161] Many women who prostitute illegally, take breaks from illegal prostitution in the legal brothels. There is a steady flow of women between illegal and legal prostitution.

A witness who requested anonymity offered the following description of an early study of HIV prevalence in the Nevada brothels. This university-sponsored study drew conclusions about HIV based on seemingly suspect data.

> In the late 1980s, Joe Conforte [an Oakland pimp who promoted prostitution in Nevada and who is now hiding in Brazil following conviction on income tax evasion] bragged about how a university in California had recently completed a study on "his girls." He stated that no one was HIV positive. I didn't believe it so I asked to see the University report. I tracked down the person who had done the study and asked how soon after the women started working at the brothel did the investigators do the first HIV test. She answered that the University had tested the women on their first day at the brothel. She stated that they retested after four months, and the test results were all negative. "But current testing requires an 18-month incubation period," I said. "Did you do follow-ups after 18 months with those same women?" She said that the University had tried but all the women in the original study were no longer at the brothel. I asked if she could find them. She said no records were kept that would allow her to track down any of the study participants to do a follow-up study.

Then I asked another question: "Who funded this study?" "It was a San Francisco organization that was set up to promote legalized prostitution," she told me. I asked, "Why would an organization wanting to promote legalized prostitution in California contract with the University to study a brothel in Nevada? She said that if they could show there was no health risk to johns it would help them get prostitution legalized in California. I then called the California Secretary of State. One of the board members of the group that funded the study was Joe Conforte [owner of the brothel where the data was collected]. A few weeks later, a friend of mine phoned the brothel where the study was done and said that she wanted to do a follow-up study on the women who were originally tested. She was told that the brothel had no idea where any of the original women were.

Amazingly, a great deal of research shows that johns frequently do not use condoms. Eighty-nine percent of Canadian customers of prostitutes refused condoms in one study.[162] Because customers the world over pay more money for not using condoms, extremely risky sex acts "can always be purchased."[163]

An economic analysis of condom use in a brothel area of eastern India found that when women insisted on condoms, they were paid 66%-79% less by johns.[164] This obviously has devastating health consequences for women since the greater the economic coercion into prostitution, the more she is likely to submit to sexual exploitation with no HIV protection.

In another study, 47% of women in U.S. prostitution stated that men expected sex without a condom; 73% reported that men offered to pay more for sex without a condom; and 45% of women said that men became abusive if they insisted that the buyers use condoms.[165]

Whether or not there are brothel policies in place for condom use, johns in Nevada request or demand that women do not use condoms.[166] Although signs are posted inside and outside Nevada brothels stating that johns are required to use condoms, there is no enforcement of this policy, other than what women attempt to do to try to save their own lives.

In Nevada as elsewhere, men bribe women in prostitution not to use condoms. We heard many such reports. Kuo observed that in the Nevada legal brothels "condoms were regularly disregarded for an additional fee and that brothel management made no attempt to discourage this." [167] Women in the brothels earn more money if they perform sex acts without condoms. One brothel, which we suspect was a location where trafficked women were prostituted, advertised that they did not use condoms for some sex acts. [168]

One woman told us that if women in her brothel tried to insist on condom use, and the john did not want to use a condom, he would phone the pimp from the bedroom, complaining that the woman "won't comply," at which point the pimp would instruct the woman to "comply."

A woman told us that she applied to prostitute at a Nevada legal brothel and she was told to begin work immediately upon her arrival in the brothel. She was hidden during a police visit because she did not have a work card or HIV screening test results.

A Nevada brothel had a medical clinic in its basement ostensibly for medical exams. However, several women stated that for routine gynecological exams, they were given a general anesthesia. They wondered whether sexual assaults had occurred while they were unconscious. "You don't have to be unconscious to have a Pap smear or be tested for STD," a woman said. "Even if they were legitimate doctors, they did illegitimate things to us."

A woman who prostituted in a legal brothel stated that HIV test results were shrouded in secrecy. She observed the following events during the 1990s when she prostituted in a Nevada brothel. After a public health nurse ushered women into the brothel office, some women left the office distraught or crying. Within an hour they would move out of the brothel. Over a period of five months, approximately thirty women departed from the brothel in this way, leading this witness to conclude that these women had contracted HIV or another STD from johns during that period of time.

A longtime survivor of Nevada brothel prostitution reported that she was driven from a legal brothel to a medical clinic in a nearby town for

STD and HIV testing. The clinic was eventually exposed as an entirely counterfeit operation. The clinic staff's degrees, licenses, and medication prescriptions were all illegitimate. This was not reported in the media. She was deeply shaken, since she assumed that her test results had been falsified.[169]

Nevada Legal Prostitution and non-HIV-related adverse physical health effects

There are a multitude of adverse physical health effects that result from all prostitution including legal brothel prostitution. One woman told us that she had "general body pain from sex being so hard on the body." A number of women had eating disorders. The 45 women we interviewed also described prostitution-related liver problems, kidney stones, joint pain, fibromyalgia, neck injuries, abscessed breasts, acne, hypothyroid, a broken knee that never properly healed, hepatitis B, and heart problems. In Nevada and also in New Zealand where prostitution is legal, we heard accounts of pimps who coerced women into continuing to service johns even when the women were bleeding from multiple sexual assaults. In both countries, women continued to hemorrhage rather than be fired.[170]

We interviewed many women in the legal brothels who were mildly intoxicated. Women tend to increase their use of drugs and alcohol when they are in prostitution because drugs and alcohol facilitate the psychological dissociation necessary to survive prostitution. Substance abuse also functions as an analgesic for physical injuries from the violence in prostitution. While the alcohol and drugs anesthetize their physical and emotional pain, the use of these substances puts the women at greater risk for harm because they are less likely to be able to defend themselves while intoxicated. Thirty one percent of the women we interviewed in the brothels felt that they should cut down on their use of drugs, and 34% felt that they should decrease their alcohol use. These are generally accepted definitions of a substance abuse problem. Although official brothel rules prohibited drug use, there was widespread use of marijuana, amphetamines, cocaine, and alcohol abuse.[171] Other researchers have reported that almost any kind of legal drug the women ask for is dispensed via prescription from physicians associated with the brothels.[172]

Viagra and other erectile dysfunction medications have probably increased the danger of cervical cancer for women in prostitution. Women uniformly fear and dislike these drugs because they are associated with longer duration of penetration. On the other hand, one pimp chortled that his business increased by 20% after the introduction of Viagra.[173] One wonders if cervical dysplasia also increased by 20%. An assistant woman pimp proudly told me that the physician who worked in her brothel was treating a currently prostituting woman for cervical cancer. Many of our interviewees expressed worry about cervical dysplasias and cancers. Several women asked us if there were any way to protect themselves from damage to and diseases of the cervix.

Stigma of prostitution despite legalization

We heard many reports of physical and verbal sexual harassment by health personnel associated with the brothels. A brothel pimp described one physician as a "dirty old man" who was married to a former prostitute. He continued to sexually harass many women in prostitution. The pimp discouraged women from reporting the man to a medical examining board because the "dirty old man" was the only physician in the area who was willing to perform the health checks on women who applied to prostitute at the brothel. Other physicians did not want the women in their waiting rooms, stating it was bad for business to be known as a doctor who treated prostitutes.[174]

Despite its legalization, prostitution is strongly stigmatized in Nevada. Women in Nevada brothel prostitution are considered social outcasts, unclean or immoral. This social ostracism results in profound isolation that prevents the women from forming supportive relationships with anyone outside the brothel.[175]

The social stigma of Nevada brothel prostitution can be seen in everyday conversation, in social practices, and in legal and illegal practices that isolate the women in the brothels from the rest of the community.

Prostitutes are deliberately segregated from the communities in which the brothels are located. In Winnemucca for example, police and pimps collaboratively prohibit women from associating with men after their brothel shifts. Prostituting women are not permitted in downtown Winnemucca after 5pm. When women prostitute at a Winnemucca brothel, their children and husbands are not permitted to live in town. None of these practices is legal.[176] As a result of this toxic prejudice, women in the brothels live double lives. They hide their prostitution, instead telling their families that they are cocktail waitresses, flight attendants. Even being a drug dealer was preferable to being a prostitute.[177]

There are negative legal and economic consequences to prostituting in Nevada. In order to obtain a permit to work in a brothel, a woman must register with the county sheriff which means that her fingerprints are sent to the FBI for a permanent record of her prostitution. If she tells an insurance company that she is prostituting, she is unlikely to be able to obtain health insurance. She is likely to be discriminated against regarding housing or future employment, and if she has a child, she could be vulnerable to legal accusations of being an unfit mother. She may not be allowed to travel to some countries where prostitutes are not permitted entry.[178]

Ironically, the United States asks on its immigration forms whether or not the applicant has ever been a prostitute. Answering yes appears to be one of the reasons to deny the right to US citizenship.

3

Pimp Subjugation of Women by Mind Control

Harvey Schwartz, Jody Williams,
and Melissa Farley

As psychiatrist Judith Herman pointed out in her 1992 landmark book on trauma, most people have no understanding of the psychological changes that take place when people are under conditions of prolonged captivity whether the captivity is overt or covert. U.S. sociologist Kathleen Barry and also Norwegian researchers Cecilie Hoigard and Liv Finstad have documented the ways that pimps control women using a process of violent 'seasoning.' Nonetheless, the emotional and mental control that pimps have over those in prostitution, including the dynamics of long term captivity are still poorly understood.[179]

Herman suggested that because people don't understand what it's like to experience prolonged terror and subjugation, they simply assume that victims of these atrocities choose to be in those situations or that they have an inferior moral character or both. The purpose of this chapter is to offer the reader a better understanding of the psychological control used by pimps over prostituted women. This chapter focuses on the *techniques used by the perpetrators* to control victims and to sustain ongoing economic and psychological exploitation of them.

Systematically and according to well-known methods, pimps psychologically traumatize women in order to establish specific types of control over them. Pimps assume psychological, biological, social, and economic control over the lives of the women they sell to johns through the use of chronic terror, cunning use of various aspects of captivity, and isolation from others who might offer support and validation. In addition they employ starvation, sleep deprivation, protein deprivation, conditioned physiologic hyperarousal, unexpected sexual violence, and learned helplessness. This set of strategies is insidiously seasoned with intermittent rewards, special favors, promises of relief and, at times, tenderness, all of which create a powerful and enduring trauma bond. As a trafficking victim in the United Kingdom explained, "Sometimes I don't see the point in doing anything. It seems useless. When someone has controlled you and made decisions for you for so long, you can't do that for yourself anymore."[180]

Pimps recruit women through a calculated courtship that is like the "love-bombing" techniques used by cults.[181] The woman is isolated from loved ones and social supports. Her attachments to others are systematically undermined and ultimately destroyed by pimps. Of course this leaves the pimp as the only available source of support, protection, or validation. The ironies, paradoxes, and reversals inherent in these relational processes can be mindboggling Parents and family members may be presented to "the recruit" as the enemy. Her children may be threatened. Society is held in contempt. Subversion of the social order becomes one of many forms of "pulling-one-over" by which the pimp feeds his inflated sense of self.

The pimp's worldview is slowly but systematically downloaded into the psyche of the prostitute. As in cults, the pimp and his "community" ultimately replace any prior social system. Sadly, victims may feel relief from both anxiety and depression when the old and shattered world view is replaced with another system - regardless of its perversity.

Whenever a victim of a totalitarian system becomes aware of elements of her/his own psychological reward or gain from adaptation to the perpetrating system, the survivor's feelings of self-betrayal, guilt, and shame can run so deep that she/he may feel unredeemable, unworthy, and deserving of all the misery she has endured.

Like other systems of subjugation, coercive control by pimps always contains the threat of violence, and the violence is periodically inflicted under conditions that maximize its effects: unpredictability, high intensity which is overwhelming to the senses and specifically activating of the victim's specific fears, aversions, and anxieties. Whenever the violence is not inflicted on her, the woman in prostitution is immensely grateful. This coerced gratitude is an essential element of the prostitute-pimp dynamic.

The overwhelming shame generated by pimps' mind control is a major barrier to recovery and healing. Confronting the shame may be so overwhelming, it may feel so isolating and insurmountable, that it makes it impossible for women to escape systems of prostitution

It is essential to remember that every aspect of the prostituted woman's body is under the control of the pimp who manages and handles her. The pimp's total control over women in prostitution includes what she wears, when and where she can sleep, and what and how much she can eat, if and how much emergency medical care she receives, even how much air and light she is allowed to have. For example in a Nevada legal brothel where women were kept in captivity, a pimp starved women if he decided they were too fat. Women in prostitution are told by pimps when they can use a toilet and whether they can or cannot use menstrual supplies. One woman who had prostituted in a legal brothel described them as "pussy penitentiaries" as in: the pimps owned her and monitored and controlled every aspect of her sexuality.

Gradually, the perpetrator succeeds in obtaining her total subordination to him. Eventually, after he has established a significant level of control, the woman trapped in prostitution will experience any use of her own initiative, her own autonomy, or even her own critical thinking, as insubordination.[182] She is bonded to him by the traumatic violence he subjects her to and the periodic indulgences he bestows on her. The complex psychobiology of trauma, attachment, and survival (brilliantly manipulated by the pimp) leaves the prostitute-victim ensnared by her own adaptation responses. Outsiders see them as partners rather than dominator and subordinate. This misinterpretation of their relationship further cements her bond to him.

The pimp is an expert at manipulating and profiting from all forms of social complicity. Where prostitution is concerned, society's readily available projections of its own failures and hypocrisies, not to mention collective forms of sexual guilt and shame projected onto those in prostitution makes the element of social complicity extremely easy for pimps to exploit.

Like military torturers, government intelligence agencies, and both mainstream and criminal/satanic cults, pimps use specific techniques of brainwashing and mind control to establish dominance and allegiance. Brainwashing is:

> Any systematic effort to instill certain attitudes and beliefs in a person's mind against her will while simultaneously inducing doubt and fear about her own sense of reality, agency, and power.[183]

Brainwashing is most effective when the victim's "cooperation" – albeit coerced, seduced, or deceptively attained -- in her own subjugation is an integral part of the picture. The ensuing shame and confusion further degrade the victim into a malleable, compliant servant of the pimp's agenda. Furthermore, the specific attitudes instilled by pimp or torturer are often completely opposed to her previous value system. For example, where she previously thought that sex with anonymous men was repugnant, she is brainwashed into thinking instead that sex with ten anonymous strangers a day is progressive, "making easy money with what you got," and a way to get ahead in the world. Redemptive spiritual beliefs may be replaced with a philosophy of nihilism and survival of the fittest. Love of family, loyalty, and duty to one's children are replaced with devotion to the pimp and his social system. The pimp/family/cult is set up in opposition to the rest of the world which is mocked for its innocence, ignorance or parallel corruptions.

Some of the specific brainwashing techniques used in any informal or formal system to control people - including systems of prostitution and sex trafficking – are:[184]

Social isolation

Women and girls are deliberately kept away from others who might offer them support and assistance. This enforced isolation is apparent in the physically remote locations of the Nevada legal brothels. Women in prostitution may also be systematically moved from one location to another not only to escape law enforcement but also to disorient them and to keep them from establishing friendships. One pimp cut off a girl's emotional support from her family by giving money to her family. The bribe to the family facilitated the pimp's control by encouraging their daughter to stay with him.

In the case of transnational trafficking, social isolation is the primary strategy of control and is easily established due to language barriers, cultural and geographic disorientation and confusion, and the confiscation of passports.[185]

Sensory deprivation/torture

A woman who was prostituted in Hawaii described being tied up and having water dripped on her forehead for more than 12 hours as a method of punishment by a gang member who pimped her.

Women and children who are being prepared for and used in prostitution are often locked up for long periods of time in windowless rooms to keep them disoriented about time and place, deprived of sunlight, and more vulnerable to the pimp's influence.

Sometimes the social isolation is such a profoundly stressful and disorganizing event that the victim will acquiesce to any form of contact, even rape, in order to be released. This "cooperation" or coerced choice is then used by the pimp to confuse her about her own motivations, and to strengthen his subjugation of her will. In extreme cases sensory deprivation and torture are used to create highly compartmentalized "prostitute personalities" in chronically traumatized and dissociative women and children.

Deliberately induced exhaustion and physical debilitation

In order to establish control or inflict punishment, pimps may refuse to provide food or water. Protein deprivation is a classic technique used in cult, military, and captivity settings to weaken victims and to impair their critical thinking abilities. According to survivor testimonies, during large conventions in Las Vegas, women who are transported from other U.S. states or from other countries are expected to service so many men a day that they become severely debilitated and sometimes permanently injured.

Complaints of fatigue or overwhelm are met with further brutality, loss of "benefits" and/or threats of harm, longer bondage, or death.

Threats to oneself and to one's family

In the United States and elsewhere, pimps commonly threaten to expose the woman who is prostituting to her family. Pimps deliberately exploit the shame associated with prostitution. Stating that she was not the type of pimp who "enforces the rules with a baseball bat," a New York pimp explained that when one of "her girls" failed to show up for work all day, necessitating cancellation of thousands of dollars worth of escort prostitution appointments, she knew how to make the woman "want to commit suicide by the end of the weekend." In one instance, this pimp telephoned the young woman's boyfriend, informing him that his longtime girlfriend was a prostitute, and offered to send a CD-ROM of pornography of the young woman to her family.[186] Pornography is often used for blackmail and to deepen the victim's complicity in her own psychological slavery by establishing her identity as a prostitute.

When women are trafficked between countries (rather than between U.S. states), they are threatened with harm to the children, family members, and loved ones left behind. They understand that the same social matrix and networks that led them into the traffickers hands in the first place still operate in their home towns. They also know that their heartfelt duty to protect loved ones from harm now requires their full cooperation with pimps who are part of the trafficking network, who are in close connection with the original pimp/recruiter.

Occasional reprieves and indulgences

Vacations or gifts may be demonstrations that the pimp is all-powerful, all-knowing and all-giving. As Jayne from Nevada explained, " the pimp takes them to nice restaurants, lets them go get her nails and feet done, go tanning or to a spa – lets them shop and buy nice things to make them think that he really cares. But all in all she's just spending a small portion of the money she made – on herself, and the only reason he's dong it is because the better she looks the more money she can make for him." The pimp convinces her that he is the only one who cares about her, and that he will take care of her forever. The dissociative survival strategies used by most women in prostitution may allow these seductions to become transiently credible and compelling. Those who are uneducated about the dynamics of prostitution find this phenomenon incomprehensible. The psychological and neurobiological reactions to alternating terrorism with gratuitous rewards deepens traumatic bonding and reinforces the twisted attachment system through a variable ratio schedule of reinforcement, among the most potent methods of behavioral conditioning known to social scientists.

Posturing as omnipotent

Jayne explained, "the pimp tries to make her feel like he is better than God. He is so superior and has so much knowledge and she should feel lucky just to be on his team, 'cause he wouldn't have just *anyone* be with him." Using an intense form of psychological invasion that simultaneously appeals to her need to be special to him, pimps tell their victims what and how they are feeling and how they know everything going on inside their minds and then prove it by citing predictable patterns of reaction and feelings based on years of experience in their trade. Using techniques like those of cult leaders, charismatic religious figures, and military drill sergeants, pimps manage to create an inflated sense of their own power, prestige, and importance. At the same time, the victim is rendered dependent on him, and provoked into initially small, and then ultimately grand gestures to win his approval, share his power and glory, and supplement the pimp's power with her own "accomplishments."

Degradation

Degradation is at the core of the enforced dependency, incapacitating shame, and mind control used by pimps. Women in prostitution are permitted no privacy, no dignity, and no bodily autonomy. Her most basic body functions are under the pimp's control. A second method of degradation is mental and verbal abuse. Many have underestimated the extreme harm caused by verbal abuse against those in prostitution which is profoundly emotionally debilitating and at the same time psychologically disorganizing. Verbal sadism enhances dependence and pimp control. Whatever shame existed in her psyche prior to the contact with the pimp is activated and magnified. After her stepfathers, uncles and brothers have started the conditioning process through sexual violation and sexual harassment, pimps hammer the message at her that she is so worthless, such a "filthy 'ho," that no one else would want to be in her presence. The pimp's attentions then become even more of a "gift" to the worthless, degraded victim, incurring even further false but binding gratitude.

Among other strategies, pimps convince the victim that she is incapable of managing her own money, and that he will, for example, invest it for her. Describing the process of degradation that is at the core of pimps' control of women in prostitution, Jayne clarified: "He makes her give up total control of herself to the pimp. She can't do nothing or go nowhere without daddy's permission because he has her believing he knows what's best for her and if she doesn't mind him she will get punished and of course it will be all her fault. He makes her believe that so she will accept her ass-whipping and then get up and go earn some money for daddy to make it up to him for her being so bad." Women may become so regressed in this extreme form of traumatic attachment that, like children trapped in abusive family environments, they may even deliberately provoke punishment in order to exercise some minimal form of control over the degradation and subjugation process. Or, worse, their shame-and-guilt-based need for punishment can become so intense that the pimp's sadism becomes almost a soothing relief.

Pimps are adept at simultaneously working multiple legal and illicit social systems. In one example, parents convinced a Nevada woman that she was a worthless mother, then proceeded to pimp her. The parents then took custody of the children, thereby receiving welfare payments. These pimp/parents also obtained all of this young woman's monies from prostitution in a Nevada legal brothel.

Enforcing capricious rules

Making up and enforcing capricious, nonsensical rules are additional techniques of maintaining absolute power, as in the case of a girl who was required by a pimp to lay down in a particular spot on the floor whenever he commanded, "down!" Creating and enforcing capricious rules has the combined effect of degrading, confusing, and reinforcing complicity in the victim - all in one set of strategies. Giving mixed and inconsistent messages, creating double binds, and forced choices also fall into this category. This particular method has the effect of undermining critical thinking, creating post-traumatic anticipatory dread, and eventually a severe, at times unremitting case of learned helplessness. The ensuing shame is so debilitating that efforts to break away (or at times, offers of assistance) can make it worse, leading the prostituted individual back through yet another loop of her own survival operations into the hands of the pimping system.

Deliberate creation of dissociated parts of the self who happily and willingly prostitute

Because of repeated traumatization and dissociative adaptation, the non-prostitute self is compartmentalized during preparation for and acts of prostitution. In some instances, specific parts of the body are dissociated for pain control during brutal assaults on the body that constitute the daily experience of prostitution. One young woman who was involved in filmed sadomasochistic prostitution in Las Vegas clubs reported the cameramen's/pimps' use of hypnotic techniques to teach women to tolerate extreme pain in order to reach a dissociated, pseudomystically named "subspace." The pimp/pornographers convinced vulnerable young women that they were "bondage sluts" or "porn stars" who demonstrated their strength based on how much violent abuse they could tolerate. Exploiting the women's need for validation and approval, the pimp/pornographers duped the young women into pathologically increasing their psychophysiological pain tolerance. As a result, highly coveted films of men's sadism against women in prostitution were produced at great profit.

Tragically, some confused and deeply trauma-bonded women in prostitution may actually derive pseudo-self-esteem benefits from so-called "mastery of pain."

Drugging and forced addiction

Drugging and forced addiction are effective techniques for controlling human behavior. Women may be forcibly addicted against their will, as in the case of a 14 year old whose face was held into a handful of cocaine.[187] Rohypnol, an odorless drug put into drinks, is used in the recruitment and early phase of prostitution to ensure compliance to the rape/sex of prostitution. "When they fed you, you started falling asleep," said a young woman who was drugged during transport for prostitution by Mexican and U.S. traffickers.[188]

Opiate addiction is a common form of psychic slavery also used in trafficking and prostitution. Seeking pain relief, oblivion, and any form of soothing, those trapped in prostitution fall prey to the mind-and-body-numbing and euphoric effects of various opiates and amphetamines. In the end, trafficked and prostituted women and children will be working manically to earn money for drug relief supplied by the pimp in order to cope with the web of torture and misery controlled by the pimp. This is an example of a highly efficient closed-loop economic system perfected by pimps and traffickers around the world.

Forced pregnancy

Some pimps deliberately impregnate women in order to control them. First, the adult woman is easier to control via her child. The mother is an easy target for threats, blackmail, and coerced choices of all kinds. Some survivors have described "breeding programs" for prostitution and trafficking crime syndicates.[189] By "raising his own ho," one pimp told Melissa Farley, he created the most malleable prostitutes, since he was able to control them from birth onwards. Other survivors of pseudo-spiritual, highly organized criminal organizations have reported similar "breeding programs." The "harvesting" of children from women in prostitution or from other marginalized groups of women in order to provide the next generation of "sex-workers" is a disturbing aspect of the sex industry which some find hard to believe.

Below is a description of three pimps who lived in a nice neighborhood, each with a different style of pimping. Nick, Larry, and Weldon are intergenerational pimps whose fathers and uncles were pimps before them. In their families, these mind control techniques were passed from father to son. Jody Williams, founder of Sex Workers Anonymous in Las Vegas, provided their stories, with editing and analysis by Harvey Schwartz and Melissa Farley.

Bait-Switch-Hook (Nick)

Nick is what was called a 'gorilla pimp' in the sex industry. He's a violent, take-no-prisoners type of pimp. Pimps like Nick are common in Nevada and everyplace else in the world. His targets are young women who are below average intelligence, who lack education, and most importantly, who lack fathers. Nick uses his attractive looks to recruit the women he turns out. He works as a bartender using alcohol to make first contact with his intended prey. He scans the crowd at the club, searching for the girls who are staying past closing time, an indicator that she lacks parental restrictions and where there's not likely to be a father at home waiting for her.

He spots his type of girl staring at him and sends her a drink on the house. The grooming process begins. Nick looks for a young girl with faddish makeup and hair style that tells him that what others think of her is important, that she needs to fit in. A woman who needs to be part of the in-crowd is very important information to him when seeking a new woman to groom and recruit.

Nick uses drugs and alcohol to control his women - so he can tell a lot by how she responds to his offer. If she *is* underage and chugs the drink down, he has a pretty good idea he has hit a good mark. She knows she can get alcohol from him despite her age and for free. She draws close to him after the club closes, wanting more to drink and more of him.

He only targets women who hang out after the club closes, wanting more drinks and wanting to get closer to him. Soon he is charming them with flattery, impairing their judgment and critical thinking with alcohol. Once the club is empty, he brings out the drugs. He gauges how far he can get with the woman by how quickly she accepts the offer of drugs. It's also pretty much a given that if she's still there after 2:00 a.m. there's no parent to interfere with his plans for her. Once she's intoxicated he lets her know there are more drugs at a party at another house. She's too stoned to drive so he offers to drive her. This isn't just to get her to the house but also to restrict her ability to leave the house and to impress her with his expensive car. Nick begins the prostitution recruitment process with a combination of seduction, entrapment, forced dependency, grooming, and bait-and-switch tactics.

When the girl arrives at the house she thinks she's going to have Nick all to herself. Instead she is greeted by four other women who currently 'work' for Nick. They all hand him money in a way that tells the new victim that this is a nightly ritual - as ordinary as brushing one's teeth before bed. At first, she thinks these women are going to intrude on her time with Nick. But once the women hand Nick the money, the women retreat to their rooms leaving her alone with him.

The million-dollar house is located up in the canyons of Los Angeles. The car they drive in is a Rolls Royce. Nick is draped in expensive jewelry. It's clear that Nick is making money beyond a bartender's salary. Instead of inciting fear, this ignites the young girl's fantasy of being special, rescued, and wealthy.

What she doesn't realize is that the house is rented from a john who is out of town. The Rolls Royce is from a chop shop[190] that Nick's family also runs. Nor does she realize that the jewelry is either stolen or traded for sex and therefore free to him. The fantasy of money and wealth is created to impress and recruit her. It is in fact an illusion and no different than the way that some cons rent an office suite to carry out their plans by fooling a mark into thinking they are a legitimate business. The fantasy environment provides cues that normalize wealth. Nick moves her from a bar to a home, since a home is considered safer than a bar. At that point, Nick moves on to the next stage of the recruitment/grooming process.

Once Nick demonstrates the money he receives from the other women - of course she's going to ask him questions about what's going on. He brushes her off, telling her he doesn't want to talk about it. Instead, he says, he only wants to listen to her. All of his attention is totally on her. All praise is on her. He acts like he's never seen a woman like her in his life. He can't give her enough attention.

In reality, all this attention is really his way of pumping her for information. He wants to know all about her family. What school does she go to? What are her hopes and ambitions? Does she have any friends or boyfriends? What's her sexual history? Nick makes it appear that he is just fascinated by her. Instead, he's searching for a woman that fits his target, a woman who would bring in high earnings if he got her in his stable. The ego-inflating attention for the young woman is essentially an interrogation interview for her, the potential product, by its manager and later distributor.

As he moves in to make love to her - he makes sure to know her sexual history first. He wants to know what unexplored sexual boundaries she has. Whatever she's never experienced before or whatever she's reserved only for special relationships – that's what he's going to move in towards during the sexual encounter. If she's never had anal sex, then he's going to pressure her for anal sex. It's important that he push an envelope with her sexually. If she's never given a blowjob then he's going to make sure she gives him one before the morning. He must be sure to leave his psychic mark on her and to create the template for future boundary violations encoded as excitement, adventure, or love.

If he can coerce her to do something she's uncomfortable with, then he has evidence that he can lead and control her into doing other things that she's uncomfortable with. He'll do whatever it takes - bribes or drugs that decrease her inhibitions - in order to get her to do something that's sexually beyond her comfort zone. He learns how to manipulate her into performing acts that are uncomfortable for her. He seduces her into betraying herself while at the same time thinking to herself that she is "expanding her limits" or "taking risks" or "winning her man."

Using the behavioral conditioning technique of systematic desensitization, pimps operate as both psychologist and animal trainer. Systematic desensitization is a process of gradually and systematically exposing a woman (in this case) to threatening or disturbing stimuli in increasing doses so that her startle or avoidance responses to that thing or idea will be dissolved over time. Systematic desensitization can be used to heal or to harm. The same methods that can be used to help a phobic child get past a fear of dogs so that she can have a family pet can also be used to disarm a woman's instinctive aversions to violence, bizarre rituals, or extreme and painful sexual activities.

Systematic desensitization is a gradual erasing of a person's limits so that the woman may not even know how she got past that point. The technique is unethically used by pimps in order to bypass the woman's natural defenses, and to dissolve her resistance to prostitution.

When she wakes up the next morning at first she feels like a queen in an expensive bed. Then a strange man who tells her he's Nick's driver tells her to get dressed so he can take her home. There is no note, no phone message. Nothing. She's completely and utterly cut off by this gesture of the stranger coming to take her home.

Of course she asks questions of the driver who ignores her and takes her home. This process is designed to ensnare her and lead into the next evening. If she's the right target and he worked his recruitment techniques properly then she'll show up the next night at the club. She'll want to know what happened: Where did he go? Who was the driver? The strategic rejection, abandonment, and confusion activate the trauma bonding process as the young victim is now essentially pursuing her predator. This turning of the tables creates the perfect template for all kinds of reversals which will be used to subjugate her later on.

Any other woman with a shred of self-esteem would be so put off by being treated this way the next morning that she would never speak to him again. But that's the whole point of his game: searching for prospects until he finds a woman without any self- esteem. When she shows up at the bar looking for him the next night he knows he's got himself a new girl. He tells her he has to work and to go home. She'll refuse to go. He makes fun of her 'square life' and tells her she doesn't belong in his world and to go home to her mama like a good little girl.

She will stay at the club all night waiting for him to finish work so they can talk. While sitting there she sees many other women hitting on him. It's important that she knows how desirable he is. Often these women are staged to flirt with him exclusively for her benefit. He sets it up so that she knows there's fierce competition for this man. Exploiting her narcissistic needs and vulnerabilities, the pimp escalates the willingness of the victim to do whatever it takes to win him, to please him, to "have" him with no idea that it is *she* who is being so insidiously and decisively "had."

After the club closes, he sits her down to have a heart to heart talk with her. He tells her he shouldn't have spent the night with her but he was just so taken with her that he couldn't help himself. He confesses that he's a pimp; he was born a pimp and will die a pimp. That any women he's with are paying him and that's his role in life. But she's special, he tells her. For the brief time they shared the night before, he was allowed to just 'be a man.'

He then lays out the bottom line. It isn't fair to the other women who are paying him if they to continue to see each other outside the pimp/whore relationship. At this point she begins to think that it's *her* idea: "if that's what it takes, then I'll do it." She would never agree to be pimped out simply as a hooker but she's convinced that they have something special. His manipulation appeals to her desire to rescue him. He's also betting that she's not going to want to lose the access to what she's been led to believe is a rich and famous lifestyle that includes all the drugs and booze that she wants. At this point Nick knows that the woman will agree to anything in order to see him again.

As he recruits and manipulates her into becoming a prostitute, he's also convincing her that they would have a relationship based on love and affection and not just cash as with the other women. This approach works well with fatherless women, those who feel a need to compete with other women for the attentions of a man.

He uses a technique called "future story-building." He talks about how great life could be together and how they'll use the money they make from this horrible life and put it to good use in the future. He has previously extracted from her what she most wants and he feeds that dream back to her as something that is truly possible through prostitution. Does she want a house on the beach? Sure. Does she want a cabin by the lake? Sure. Whatever her dream is, that's what they'll do once they make enough money. It's just a little short term pain for long term gain.

Besides, he tells her, the other women are the ones who will be doing most of the work. She thinks she's going to be managing the other women who are the ones who will be prostituting. Nick actually does give her this preferential treatment for a short time. He convinces her that she'll never have sex with anyone but him because he loves her too much for that. He tells her that he's just using the other women for money. He attempts to instill in her a sense of superiority over the other women. This will be the means by which he'll manipulate her in the future. She will want to remain in the top dog position as time goes on.

Coerced competition, coerced gratitude, and coerced complicity operate in a diabolical synchrony to ensnare the woman - who is vulnerable because of previous psychological injuries - in a web that she will need outside support to extricate herself from.

He brings her home to meet the other women and to get to know what the business is like. She's allowed to answer the phone while the other girls go on calls. He declares that she's too good to go on any of these calls - at first.

During a two-week indoctrination period she's made to believe that Nick adores her. He takes her shopping to buy the right clothes for the job. She buys jewelry, makeup, shoes, no matter how expensive. There's another reason for this as well: once she goes home with these items someone in her family is going to ask questions. The shopping spree is aimed at flushing out anyone who might be an objector, perhaps a mother or spouse. The bigger the diamond, the quicker an objector in her life is going to be flushed out. This strategy also works to either seduce the family into cooperating or to pit the women against her family. Sadly, she fails to see the pimp's hypocrisy in taking one woman's money to buy another woman gifts.

After about a week, any objectors will come forward. They demand that she stop seeing this man before it's too late. She on the other hand believes that she's going to be on the inside of a hip new crowd with lots of drugs, money, jewelry and fast cars. She feels like Cinderella at the ball with all the other women jealous of her. There's no way she'll give that up.

After the objectors fail to convince her that Nick is dangerous, they may then may go directly to Nick to persuade him to leave her alone. At this point Nick will do whatever he can to get the girl kicked out of where she's staying. She will then have nowhere to go except his house.

Secretly the girl is hoping that she will get kicked out so that she'll get closer to Nick. Then, she thinks, she'll be queen of the expensive house and underground world. She sees the choice as between a big house and a Rolls Royce or going home to a one-bedroom apartment with her mom yelling at her about home work. Nick has done everything he can to build a fantasy world that is far better than the real world at home.

Nick targets young uneducated women, most of whom are living with someone who cares for them. Leaving home means that she will be dependent on Nick. Once they've left home they won't want to lose face by going back. After confrontations with objectors, she'll be more determined than ever to make it work out with Nick. Nick has now placed her in the position of defending his lifestyle to her family. He never shows his real hand until she's left home and is now totally dependent on him.

Once he gets her into the house and dependent on him, the rules change. She's no longer an outsider. Now she's part of the family system in the house. Nick explains to her that he's not alone. There are other women and other family members and a generations-long system in place that she'll have to adapt to in order to stay with him. She's told that she is one of the family now.

She's told that she must learn to adapt or the family will reject her. One of the family rules is to do what's best for the family even if it is counter to her self-interest. It begins with little things like making dinner for the whole family or doing everyone's laundry. Everything is now geared towards her understanding that this isn't just her and Nick anymore. She's now responsible for everyone's well being. Although the system is pitched to the new recruit as some form of collective, in reality it is an elaborate sex-plantation system replete with overseers, field hands, house servants and the like.

This is when she learns that when she makes a mistake other people are hurt. If she burns toast at breakfast, one of the other women is beaten because she should have been helping her with the toaster or have made the toast herself. If the clothes aren't ironed properly - one of the other women gets yelled at. This new girl is learning that when she makes a mistake, everyone suffers. She starts to lose her sense of self and starts identifying with the group. If someone doesn't bring home enough money, everyone is punished. The military uses this same strategy to break down individualism and to build unit cohesiveness. The Nazis took this to an extreme in the concentration camps by murdering dozens of inmates whenever there was any escape. This placed extreme pressure on the inmates to protect each other's lives by "staying put" and cooperating.

Nick doesn't aim his anger towards the new recruit yet. He still wants her to feel they are a team and that she's better than the other women who are taking his abuse and handing him money. This is an elaborate play written, conceived, and directed by the pimp in which he is also playing the dominant role while seducing the new recruit into believing it is she who will become a "star."

Then comes the pseudo-crisis engineered by Nick. Maybe the car has broken down or the ice machine at the club malfunctions. Maybe his mother needs an operation. An imaginary crisis is created so that a lot of money is needed in a short period of time. At this point he starts to withdraw his attention. He convinces her it's because of the crisis and raising the cash is the only thing on his mind. He has to focus on the other women now because they are the ones bringing in the money he so desperately needs. She senses she's losing his attention but at the same time she wants to please and rescue him. Her attachment and rejection emotional wiring are being systematically short-circuited.

A john magically appears on the scene who happens to have exactly the amount of money that Nick needs and he wants a new girl. She is convinced that the only solution to the crisis is to actually go on the call along with one of the other women. She's told that she won't be left alone, she'll be with one of the other women who has offered to take care of her and make sure nothing happens to her. She has no idea that the john was selected to be unusually attractive and generous. She's presented with a very skewed picture of what johns are really like by this man who is staged for her benefit.

Nick's attention has been withdrawn from her during the pseudocrisis he's created. She returns from the easy date with the john where a false world and fake client were introduced to her. As Nick lavishes attention and gratitude on her, she thinks to herself that it wasn't too bad. She's officially a member of the family. There's even a group celebration. Everyone welcomes her now that she has taken one for the team. This is an initiation of sorts in which the victim believes that she is now willingly joining the family. She does not understand or even see the complex system organized to exploit her.

Initiations are used in almost every culture, from the underworld to the corporate world, from professional sports to military training, from mainstream religion to satanic cults. They strengthen emotional bonds among group members, derail critical thinking, and cultivate an 'insider' status with feelings of specialness and superiority which sustain and at the same time restrain the individual when autonomous or rebellious strivings emerge.

The next morning the new recruit is in for a rude awakening. As the women prepare to go to work in the sex industry, she's told to get ready for work with them too. She resists because she feels that she is not 'like that.' She appeals to Nick at which point Nick shows his real face. He tells her she's already crossed the line: she's now a whore just like the others. He tells her that she's now 'damaged goods.' Now, just like the others she will only receive his attention if she makes money. If she angrily resists or refuses, the tables begin to turn against her.

The woman who makes the most money at the end of the night is now announced as the top girl for the night. Nick now invites that girl to share his bed for the night with him - pushing the new girl out of his bed and into the room with the others. Now the rules are: whoever makes the most money gets to spend the night with him.

She has now been demoted from sleeping with Nick to sleeping in the room with the other girls. She is now set up to work hard for something that was once freely given. She is by this time so confused and deceived that she may not even critically question her own desires or motivations. This is when the other women start coaching her about what's really going on, what Nick is really like, and what the rules of the game now are for her. It's a demotion and every fiber of her wants to go back to being on top so she listens. Her competitive and ambitious personality traits are now activated. These women are becoming her sisters and best friends as well as her rivals. She also has nowhere else to go now so she listens. She has to find a way to get back on top and make the best of her new home so she listens. The horror stories are revealed. She learns about Nick's extreme violence, hears about women he's hurt before, and what his boundaries are so she doesn't get hurt. The sister/rival splits further the confusion and trauma-bonding system that holds the entire prostitution matrix together.

From one point of view, it looks like a big slumber party. But what's really happening is that the "new family member" is being taught that Nick is not someone to mess with or be disobeyed. Or else. For a while this intimidation works. She is afraid to confront or challenge him in any way. She still does not realize what she has fallen into and she still may think she can master the situation. All the better for the pimp since her efforts at mastery will all be turned against her and exploited by the pimp. She has been so effectively set up that all her efforts will work against her and for his cause.

In this system, Nick does not make direct threats. That would be less effective mind control. Instead the other women are the ones scaring her with stories of his past behavior. They are unwittingly indoctrinating her. If arrested, Nick can claim that he's never made a threat or coerced her. The pimping system that controls her has become highly efficient: it is self-reinforcing, self-administering, self-monitoring, and self-controlling.

With time the new woman wonders if he's really as dangerous as they say. Pimps talk about how a girl will provoke violence by challenging rules they know will be met with violence. The truth is that she starts to think about leaving this life. She gets scared thinking about the horrible things that might happen to her if she tries to escape. She wonders if the talk is just hot air and that he'll actually do nothing to her if she leaves or refuses to give him all her money. Her autonomy has not yet been fully broken. She is still calculating, strategizing, hoping. She has no idea that the web she is entangled in has so successfully ensnared her that any movement she makes will bind her further and deeper into the web.

She will inevitably challenge him in some way in order to test the waters. This is when he has to establish that the stories are true. He either punishes her directly or he strategically punishes another girl in order to set an example for maximum effect. But he also holds out hope that things will go back to the way they were. He establishes a point system of demerits and rewards.

The system would not work if it were as simple as whoever makes the most money gets all of Nick's attention. In that case the prettiest or most experienced girl would always earn the most money. The others would give up and then Nick would only have one girl. Even if that girl made a lot of money - at any point she could get sick, arrested, hurt or even killed. He must always have at least one back-up girl in play.

It's a system with checks and balances. If the top-earning girl had no competition she might stop working and then she'd have control over Nick. What keeps her going has to be the hope that she's going to make the most money that night. Nick has to know when the other women are thinking of giving up the competition for him and he has to know when the top earning woman starts to get cocky because she has him all to herself. His savvy insights about human nature – sociopathic wisdom - allow him to watch his crew like a hawk or a guard dog sensitive for anything - any movement or subtle changes - that seems out of line or threatening to his status quo.

When this threat or change happens he provokes a fight. He might accuse the top earner of holding back the money or of talking to another man. The fight is arbitrary; it's about something that she's not doing or that she has no control over. She has no idea why he's angry at her. He deliberately stuns her with his sudden rage and the incomprehensibility of the charges against her. She is caught in a double bind. If she fights against his accusations she is proving him right, and if she accepts them she is proving him right.

The intensity of his rage creates shock and terror in her. In this adrenalin-charged state, everything he says to her goes directly into her subconscious as hypnotic and posthypnotic suggestions. She is left vulnerable to his commands, to his ever-shifting versions of reality, to his definition of who she is, to his withdrawal of attention and validation.

Nick kept women working for him for many years using this cult/family system. Military-like loyalty binds a group of victims together, each vying for her piece of the imaginary pie, feeding the system and the pimp at their own expense while seeking deeper comfort and security inside the system that is consuming them like any other malignant and ruthless profit-based enterprise.

Long-term Grooming (Weldon)

Nick's brother Weldon used a different approach to pimping women. While earning his Master's degree at UCLA he searched for women with whom he could build a complex and intimate pimping relationship.

Weldon searched for strong, ambitious African American women. He knew they were discriminated against within the primarily white college crowd of the times. He knew that because of racism, their economic options were more limited than those of their white sisters. Unlike his brother Nick, Weldon did not have to establish the vehicle of isolation, he hunted and profited from a group of intelligent women who were already marginalized and disenfranchised.

Weldon targeted women who wanted to get ahead in life, those who wanted an education but who felt that the deck was stacked against them. Weldon would look for attractive, intelligent women who were supporting a dependent - a child, a disabled mother, a retarded brother, etc. He looked for someone who would be working two jobs while going to school full-time. Unlike others who preyed upon the lost, immature, and dysfunctional, or "good girls" who wanted to "be bad," Weldon's niche targets had to have competence, dedication, solid values, reasonable morals and a willingness to work themselves to the bone for loved ones, for a good cause, and/or for their own ambitions. Practical, grounded, hard-workers were Weldon's specialty.

Unlike Nick who counted on mind-altering substances to further his aims, Weldon screened his targets for any drug or alcohol use. He avoided promiscuous women. He wanted focused women with clear heads who were strictly interested in the money. Above all, Weldon was a businessman and the inflation of power came through money and not through narcissistic preoccupations, extreme dominance, or self-aggrandizing displays. Without the use of addictive substances, powerplays, social pressure and elaborate romantic schemes, Weldon's sophisticated form of grooming and mind control was even more insidious, potent, and far-reaching than Nick's.

We all say we'd never kill anyone but the truth is that it's all about context. If you had to kill someone to save your child then the answer about what you might do would be different. Weldon did everything through the use of logic and words alone. He knew how to move a person along their edge towards participation in his schemes at such a slow and systematic rate, using empathy and camaraderie, that the woman-victim was unaware of the distance she had been "taken" from her own starting point.

Before long she would be capable of many actions she had previously thought impossible. Of course, each act solidified Weldon's base and brought her further into his subtle but pervasive control. Unlike Nick who focused on speedily manipulating women's *behavior,* Weldon's more subtle use of systematic desensitization was aimed at gradual shifts in the woman's *attitudes and thinking.* Weldon's recruits to prostitution were eventually operating independently as if their systems of motivations and goals originated from within them. The end result however was the same. Both pimps shared the goal of controlling and exploiting another person.

Weldon would become an understanding person who offered emotional support for the tough road ahead. Weldon used himself as a lure. He was an intelligent, mature, African-American man who provided emotional support for these over worked and very stressed women's lives. The fact that there were few men like him helped him in recruiting African American women for prostitution. The lack of competition for his particular niche worked well for Weldon. From keen observational skills developed when he was young, Weldon understood all too well the race-and gender-based vulnerabilities of the women he grew up with. He knew their hopes, their dreams, the angers and frustrations with men, and he knew ALL of their vulnerabilities.

Weldon wouldn't consider the use of physical violence against his women. Weldon said, you can't intimidate a truly intelligent woman, you can only 'motivate' her. Motivation is exactly what he would do. Anything but romantic, Weldon's meetings with these women were more like sales meetings. He exploited the reality of racism and used that against her. He used the racism of US culture as a manipulative device to convince her that the odds were slim that she'd find an African American man to marry and care for her. He outlined to these women how working in the sex industry while using him as a mentor and more of a friend than lover was going to get them ahead and also secure in life. Convincing a woman that he had her best interests in mind, Weldon laid out his five-year plan which included his managing their earnings from prostitution as they earned a degree and he helped them achieve financial security for a lifetime. He used chalkboards to make his sales pitch to women who were in school; it was a soft programming.

Weldon's techniques blended variations of the corporate world, the non-profit world, the church/ministerial world, drug cartels, Soviet five-year plans, the modern business-oriented "Mafia" and "the academy" in a startlingly powerful blend of tactics. He could probably teach a thing or two to business and political leaders about how to maximize profits while keeping employees happy.

Weldon disagreed with Nick's techniques. While Nick would turn out his women in less than two weeks, Weldon took at least six months to build a relationships with a woman before manipulating her into acquiescing to prostitution with himself the pimp. Of course he never called it pimping. Women were intensely loyal to Weldon, some prostituting for him for 20 years. Weldon felt that Nick's tactics used women up and burned them out, eventually causing them to leave. Weldon preferred to avoid the turnover and the threat of an angry woman who was injured by violence going either to the police or to the competition. Weldon was a long-distance runner pimp while Nick can be seen as a sprinter. Weldon made long term investments while Nick used what he could and threw away the rest. The contrast between these two brothers mirrors a similar split in plantation, factory, and business management in the so-called "legitimate" world.

Weldon felt that his method fostered true as opposed to coerced loyalty, long term income and life-long relationships. His women did not get arrested in part because they never prostituted in the street. Weldon didn't accrue medical bills or lawyers' fees the way Nick and other pimps did. Weldon cared for his "property," was rarely impulsive, and invested in his women economically and psychologically like they were thoroughbred race horses trained and conditioned for stakes racing.

There was no arguing with these women about the exploitation in their relationships with Weldon. Conflicts and dramas were kept to a minimum since he was only supporting, guiding, and managing them. He wasn't taking their money - he was investing it in their future. In five years the women would have university degrees and money in the bank with bills paid off. They didn't see Weldon as cheating on them with other women. Instead of a lover he was a long term family friend. Weldon never raised his voice or hit the women. The absence of drugs, romance, and violence is an essential distinguishing characteristic of this form of pimping.

Weldon collected the women's prostitution monies and placed it equities, bonds, or annuities with a portfolio manager who set up a comfortable monthly budget.

Weldon was so logical that the women really believed that to leave him or to stray from their long-term plan would be crazy - not the other way around. Weldon had miraculously merged his own self-interest with the self-interest of his women so there was little need to subjugate their critical thinking in traditional cult-like ways. His form of seduction was so quintessentially pragmatic and in line with "the American dream machine" that everything fell easily into line, keeping him in the role of benevolent leader-brother-father-manager figure for extremely long periods of time.

But Weldon's women had no long-term relationships. They put off having children and many were sterilized. Weldon wove a fantasy that became more and more who they were. They stopped letting anything into their lives but work. No shopping, no friends, no romance - because this meant time and energy away from the master plan. Normal recreation was viewed as a waste of valuable time. These women's normal emotions were compartmentalized out of the way and they were programmed to prostitute. This scenario is not unlike business training and motivational movements that ignite hypomanic behavior, encouraging compartmentalizing feelings and conflict in the service of the profit motive, greed, or empire-building.

Weldon used the women's lack of choices against them, convincing them relentlessly and logically that only he could help them get ahead. Although his approach was not violent, he overwhelmed the women with the daily grind of prostitution, keeping them socially isolated and ultimately, like Nick, programming the women into money-generating automatons. Over time just like Nick's women, Weldon's women internalized the view of themselves as hard-working whores. Identity alteration is a final part of the programmed pimp-prostitute relationship, whatever the methods used to establish it. Because of the intensity of Weldon's programming, the women saw themselves as different from others; special women who had chosen an alternative road that normal people wouldn't understand. The sex industry wasn't just a job for them. It became who they were for the rest of their lives. Combining specialness, "outsiderness," and ambition, Weldon successfully managed the labor and identities of a reasonably sized crew of women who prostituted for long periods of time.

Nick created an environment where prostitution was normal and it seemed abnormal not to be prostituting. Weldon did the same thing via a carefully constructed mix of intelligent chains of lies and dustings of truth, ensnaring economically vulnerable women who had been harmed by racism and sexism. The larger social context of the media-generated American dream bolstered Weldon's efforts. The salve of material gain was used to soothe and secure women the way that other pimps use drugs, seduction, and violence.

Degradation and Trafficking (Larry)

Larry was Nick and Weldon's middle brother. Larry had his own pimping style. He liked to own his women and he did that by buying them young. He made sure that women would not be able to function in the world without him if they escaped. Larry is as much a sex trafficker as is a Russian trafficker who transports a Romanian girl for sale to prostitute in Amsterdam.

Tina's mother was a junkie who sold her to a traveling circus when she was eight years old. The carnies used her for sex and for various cons and for servant work around the camp. It is hard for most people to realize that parents throughout the world, including in the developed world, are willing to sell their children to sex traffickers for a variety of reasons and under a number of pretenses. However in every society there are people who are prepared to sell their children and there are those who are prepared to buy: pimps and traffickers who know how to make the connections.

When she was twelve years old, Larry bought Tina from the circus for $2000. She was relieved to be owned by one man instead of the entire circus staff. She also knew that nobody would believe her if she told them what was being done to her. This relief at single ownership is similar to the relief prison rape victims feel when they only have to service one inmate instead of enduring ongoing gang rapes from the entire prison population. The owner protects his 'property' in exchange for sole use. The relief is a kind of coerced gratitude that functions as yet another means of psychological control based on traumatic bonding. The person in captivity is grateful to the perpetrator that the harm is not worse - that it's only one person's daily rape as opposed to twenty peoples' daily rape. This gratitude is difficult to grasp if one has not been in a hostage or captive situation.

Like women trafficked from outside the United States, Tina had no birth certificate or other identifying documents. The only people Tina ever saw were tricks – and they never asked her why she wasn't in school. Larry rented an apartment for Tina and gave her what she needed to continually service tricks. Although she earned about $10,000 a week, Tina had no makeup because Larry said it was a waste of his money. Larry deliberately kept Tina illiterate as a means of keeping her dependent on him. She couldn't write a check or read utility bills. The use of forced illiteracy to subjugate human beings is as old as slavery itself and was, as most people now understand, an essential and vehemently enforced aspect of North American slavery. Lack of knowledge and access to information as a tool of social control is also practiced by many totalitarian regimes and other cultic groups. Stopping critical thinking from developing altogether by limiting access to what would facilitate its development is far easier than crushing critical thinking later on.

I also saw Larry buy Christine, a scruffy girl who looked to be thirteen or fourteen years old. Christine's mother didn't feel like supporting her and Larry had promised her mother that he would teach her the business, that Christine would learn a good trade while living in a nice house with pretty clothes to wear. Christine's mother traded her to Larry in exchange for a car. Difficult as it may be for some people to believe, it is not that difficult for a shrewd pimp in a Western industrialized nation, wise in the ways of marginal parent-child relationships, to develop a stable of young girls in this manner.

Larry sent Christine to school every day because she was in school at the time he bought her. So that no questions would be asked by the school system, Larry beat Christine if she didn't receive good grades. When she turned sixteen and was legally permitted to drop out of school, Larry celebrated her independence from the school system with a party. Pimps coerce victim compliance by using the system against itself, playing by the rules in order to break the rules, and camouflaging criminal behavior behind "normal" appearances.

Christine knew her mother didn't want her and had traded her for a car. She felt that living with Larry was better than a foster home or the streets. Nick and Weldon, like other sexually and physically abusive men, used the victim's gratitude that it could be much worse - as a means of cementing traumatic bonds.

Larry's system was straightforward enough: work 24/7 turning tricks any way you can, then turn all the money over to him without question when he'd come by on his nightly pickup rounds. Other rules included do not talk to Larry about your life, stay away from all Black men since they could be competitor pimps, no intercourse without a condom with any man but Larry. If anyone disobeyed Larry in any way, she would be brutally beaten. Whoever made the most money per week was allowed to come spend the night with him at his place, ride in the Rolls Royce, and sleep on satin sheets with him that night. These pimping systems have its own rewards and punishments, forming a unique subculture or alternative family system. Most of those involved "buy into the game."

Since Larry did not allow his girls to keep any cash, if a girl wanted something that Larry didn't buy, she had to find a john who would exchange it for sex. If she wanted clothes, she had to find a john. Larry's women even had a list of gas stations that would trade blowjobs for a tank of gasoline. Depriving the victims of currency is another vehicle to thwart autonomy. It forces the prostituted woman deeper into her "chosen line of work" so that she has a sense of pseudoautonomy and "nothing to complain about." Once enforced, endured, and practiced - even the most bizarre sets of rules begin to make sense.

After one of Larry's beatings left Tina black and blue, I helped her with makeup, new clothes and a new hairstyle. Larry was furious. He threw Tina's new clothes out, destroyed her makeup and took scissors to destroy her new haircut. Upset and angry at all this, I confronted Larry.

He explained to me that the way he kept control of his women was by "keeping them low." He *wanted* them to be insulted by johns for looking tired or ugly or old and compared by the johns as not as pretty as some of the others. He wanted them to be paid less than other women. He wanted them to feel 'less than' other women. His philosophy was: if you dressed up a whore and put nice things on her - pretty soon she'd start thinking she was somebody and would realize that she was a human being after all. She would soon realize that she didn't need him. If the women started feeling better about themselves, they might make friends. Larry did not want his women having any friends. He wanted them dependent on him for everything in their lives. Larry explained that he forced his girls to work after a beating because he knew that tricks would recoil in disgust and bargain down the girl's asking price. He wanted her at that moment to realize that the only person she had in the world was Larry. When I pointed out that Tina made twice as much money by looking better, Larry responded, "I'd rather have less from a beat up worn out whore then more from one who started to wake up and realize that they don't need me anymore."

Larry did not permit his women to obtain identification documents of any kind. The reason he bought his girls young was that most didn't know enough about their birth names, parents' names or even their own full names to obtain a birth certificate or a driver's license. This became clear when Christine delivered Larry's baby. The hospital asked for some identification but she didn't have any. They asked questions to help her figure out how to order her birth certificate and she could not remember her mother's maiden name or her mother's birth date.

Like Christine, Tina didn't know her birth date because when she was sold in the circus it was never celebrated. She never knew her last name because she was sold simply as 'Tiny Tina.' The lack of documentation kept women dependent on Larry just as it keeps internationally trafficked women dependent on traffickers and other organized criminals. The women felt that a beating once a month was worth it to keep a roof over their heads because they felt that they had no other options for survival. Because they were illiterate, these women could not search in a phone book for social services, welfare, shelters, or something they rarely dreamed of: going to school for job training.

Without identity or identity papers, currency, literacy, outside support or any witnesses to this form of psychological and sex slavery, and with various forms of "breeding programs" added to the entrapment, the young female victim is completely ensnared in a web no one can get her out of. In many cases, it becomes a web she may cease to consciously acknowledge that she even wishes to get out of. It may appear to the victim and to society as her own 'choice,' made of her own 'free will.'

Larry's brainwashing was conducted during sexual abuse or beatings when she was trapped, powerless, and terrorized. When a person feels terror, it produces a heightened physiological state that enables words to penetrate deeply in her being - into her unconscious mind. The pimp/trafficker's words imprint themselves on her soul as well as her mind and body. His thinking patterns become her own.

When the women joked that they were just "stupid ugly whores not good for anything else" it was not simply that they had low self-esteem. They had been programmed by Larry's techniques of physical, sexual and emotional abuse to believe that was truly who they were as people. Split off from the debilitating shame that was their everyday existence, the women found humor in "trashing" themselves and one another as a way to gain some form of minimal psychological control over the situation.

Larry was very careful not to let them have anything that might shift that degraded identity, even if it was something as simple as a gift from a friend. If a john gave a girl a present Larry would either whisk it away or destroy it right in front of them. Once a john gave Tina a cat. She began to bond with the cat, creating a makeshift litter box and using a towel as a bed. That night Larry saw the cat, grabbed it and walked out the door with it. Tina never saw it again. All forms of unconditional love - even that of animals - must now be kept at bay so that the pimp's conditioning matrix that holds the women in psychological slavery would not be threatened or disrupted.

The pimp enforces a learned helplessness that keeps women utterly dependent on him and malleable to whatever his instructions are. Wherever she turns for soothing, for friendship, the pimp is there to punish her or remove the source of comfort if it is not under his control. This futility programming eliminates resistance. Tina never tried to obtain a pet after that. When a friend brought her dog over, Tina even stopped playing with him. Internalizing Larry's training, Tina stopped even thinking of trying to get anything she knew he would disapprove of. There was no point in enrolling in a computer school if he'd trash the computer the minute he saw it. Any place, object, person, or activity that wasn't part of making money from the sex industry - was removed from her life immediately. Since any other activity would result in a beating from Larry, his women focused on prostitution and nothing else. For better or worse, dissociative absorption is a traumatic survival strategy of many survivors. The traumatized prostituted woman's natural psychological defenses actually worked to drive her deeper into the hands of her pimp.

The older Larry's women became, the heavier their despair settled on them. How could they escape? At age thirty, Larry's women had been raised like animals, constantly sexually abused, beaten, emotionally degraded, not permitted any clothes except 'ho outfits, one-dollar sandals for shoes, illiterate, with no ID, no job skills. Their chances for escape from Larry were minimal to none.

Larry began using drugs, forgot to pay Tina's rent and she was evicted. She temporarily lived in a camper with no heat. Because she was cold, a john offered her heroin to warm up and overdosed her. Tina died from the overdose. The john abandoned her when he realized she was dead. Her body was discovered some time later.

Larry was sent to prison for the sale and transport of heroin. Weldon and Nick did not want the women who didn't even have the skills to pay their rent with a money order. Christine moved into Larry's apartment with a three year-old and continued to turn tricks to pay for rent and food.

The effectiveness of pimp brainwashing is tragically apparent when women in prostitution encounter systems of law enforcement or social services. Even those who could help women escape it are unaware that prostitution is not what it seems to be on the surface. Women who appear to be "consenting upper class escorts" or "tough street whores with an attitude" are not often understood as being under pimp mind control. The sophisticated techniques that pimps use deliberately conceal the real nature of their control. Trafficked women are brainwashed/terrorized to say "I'm happy and I'm making money" and somehow police and social workers alike accept that implausible statement, even though they themselves would never feel comfortable having sex for money with 5-10 strangers a day. The fact that the person making the statement believes it herself makes it even more convincing. But that does not erase the fact that the person making the statement has been brainwashed and is more often than not currently under pimp control.

The harsh reality is that a majority of women in prostitution are under pimp terrorism and mind-control just like that in Jonestown, the Moonies, the Symbionese Liberation Army, and other cult or paramilitary groups where human beings may act happy to sell flowers by the side of the road, drink Kool Aid, pillage, plunder, and murder, or sell their bodies for any of a number of fantasied redemptive or protective possibilities offered in these deceptive subcultures. This fact is not in any way whatsoever a reflection on their worth as human beings any more than the facts of women's being raped or battered reflects on their worth as a human beings. Indeed, the mind control is a direct result of complex posttraumatic stress disorder, trauma bonding, behavior modification techniques, and the loss of critical thinking abilities that occur under prolonged captivity even when that captivity is hidden by cloaking devices.

We must look at what is behind the facade of the sex industry. Because some prostitution venues have neon lights, large parking lots, women who are attractive and who smile and accept cash, many people look no further than this thin veneer. This shallow analysis whitewashes the sexual exploitation and violence intrinsic to prostitution. Misunderstanding the origin and meaning of the smiling of women in prostitution/pornography or the smiling of the trained child soldiers butchering a village in Africa under the orders of their commander, or the smiles of Moonies - the uneducated public is easily distracted from the corruption and ruthlessness that created those smiles. Some refuse to acknowledge that even in the upscale venues there is likely the same kind of pimp brainwashing that we've described in this chapter. People don't want to see the predation that's in plain sight - the steroid-saturated and amphetaminized bouncers who guard the doors, the barbed wire around the back door, the police presence, the occasional busts of violent organized criminals. They refuse to name the taxi driver a pimp when they hand over $200 to buy a young woman to be sent to their hotel. They certainly don't name themselves predator. Instead they consider themselves to be players or hobbyists. To see such a seamlessly executed system successfully holding large numbers of human beings in captivity would change their world-views forever and disrupt core beliefs about safety, society, and human nature itself. Unfortunately, most people bargain away painful truths in order not to have to face such existential challenges.

The psychology of cults[191] and pimping/prostitution have much in common. While the cult leader exploits the follower's spiritual idealism, hunger for spiritual guidance, relief from despair, fear, loneliness, or existential dilemmas, the pimp additionally exploits economic and psychological vulnerabilities, sexual and physical abuse history, isolation, and material dreams of the victims. Both systems rely heavily on recruitment strategies, disinformation tactics, falsification of leader/pimp's biographies, isolation from dissimilar influences, mind-numbing and trance-inducing activities, techniques of intimidation and humiliation, demands for submission/surrender, and thought reform. Both cult leader and pimp have mastered the art of exploiting the universal human dependency and attachment needs. Both types of totalitarian domination create isolation from discrepant information, discourage critical thinking and foster a need to banish objectivity

Moral relativism, superiority and arrogance are common to pimp and cult leader alike. Based on their 'special understandings' of the human condition - pimps and cult leaders use persuasive justifications and seductive philosophies to dissolve victim resistance while at the same time solidifying traumatic bonds between themselves and their followers/stable members. Women in prostitution, just like cult followers, find transient refuge in the shadow of their commanders' omnipotence.

Sadly, the public's confusion over and marginalization of cult victims parallels that of prostitution. Victims are blamed and pathologized rather than understood, supported, and treated. Survivors of both cults and prostitution are maligned, ridiculed, and treated without compassionate understanding of the all-too-common human dynamic in which they have lost their freedom, dignity, and selfhood.

Oppressive criminal systems brainwash not only the victims who are trapped inside. They also create cover narratives, seductions, and distractions that promote dissociation in the general public. These cover dialogues obscure the violence that is intrinsic to prostitution. We ask ourselves, how could something that cruel, that vicious be advertised publicly as "Live Sexy Girls" right there in plain sight? As we'll see in the chapter on advertising, almost half of the people visiting Las Vegas assume that prostitution is legal although it is not. Sex industry entrepreneurs, organized crime, sexually predatory johns, and the politicians who are complicit with them together construct elaborate camouflage that obscures the nature of the harm in prostitution. They promote social amnesia.

Similarly, the Nazis constructed an elaborate camouflage in the Terezin concentration camp so that it would appear to be a summer camp for Jews rather than a death camp for them. Even though they investigated, the Red Cross was fooled. So was the rest of the world. The "duping delight"[192] of perpetrators as they pull off their hoaxes is the frosting on the cake of economic and psychological benefits obtained from their simultaneous management of victims and collaborators.

As Belgian King Leopold savaged the Congo in a bloodbath of destruction, a hundred years of colonial violence was carefully painted as a mutually beneficial trade pact between Africa and Europe. Leopold and his trading partners fabricated organizational names that were designed to camouflage the nature of the Belgian rape of the Congo and its people. One organization controlled by Leopold was the humanistic-sounding International Association of the Congo. Leopold certainly did not talk about starving porters, raped hostages, emaciated slaves, and severed hands.[193]

Slavery in the United States also had cover narratives that led to collective dissociation of what was actually going on inside the system, what made it work, and the lethality of it to the slaves involved. All intelligently designed oppressive systems use variations on these tactics. Most important is garnering the complicity of the masses through exploitation of peoples' ignorance and incredulity. This is accomplished by the deceptive and distracting use of language, by exploiting people's natural self-interest, fears, avoidances, anxieties, and by exploiting the dynamics of sacrifice that happen in human relationships and societies.

The interlocking systems of drugs, prostitution, mind control, and economic exploitation of those trapped in prostitution are mind-boggling. One snapshot alone can be deceiving. Only by taking a panoramic view of its entirety can the public become educated about this system of modern day sexual slavery.

We live in a world where we somehow accept that thousands of humans a day can die or be sacrificed from hunger and starvation, thousands can be left uneducated and unattended in poverty, and hundreds of thousands of people a day can be emotionally and physically brutalized and psychosexually enslaved in systems of prostitution and trafficking. We act as if there are no consequences to our refusal to acknowledge these sacrificed, marginalized, dehumanized people. We forget that our own humanity is compromised when anyone's humanity is compromised.

4
Johns in Legal and
Illegal Nevada Prostitution

"There must be girls, as many as the men demand," wrote Gabriel Vogliotti, quoting gangster Bugsy Siegel's business approach to Las Vegas.[194] That sales approach is still in place in Nevada. Men who arrive in Nevada to buy women in prostitution do so for the same reasons that men do everyplace else in the world: they want to purchase dominance.[195]

Johns are the designated market for trafficked women who are advertised to accommodate their preferences for the young, the vulnerable, the "exotic," the one who loves pain,[196] the woman/girl who is racially or physically different from the john, the dirty one. Legal brothels and illegal outcall and strip club prostitution provide the john with whomever his fantasy requires, ranging from "minstering angel to filthy slut."[197]

Signage inside the legal brothels offer information about what pimps think would influence johns to buy women. For example, a sign reflecting anti-government, old-west thinking:

> Best in the West
> Cathouse Tax Service
> Open 24 hours
> We Fuck the IRS

Another sign suggested to the john that if he did not have a wife who is sexually accessible, you can buy a prostitute. The original sign had read WIFI available here. The sign had been modified to read:

> WIFE's
> available
> here

A sign at another brothel admonished johns to buy and not just look, suggesting that a man is a wimp if he hesitates to buy a prostituted woman, and that he is a loser if he considers masturbation as an alternative to the use of women in prostitution:

> He Who Hesitates, Masturbates

Peoples' acceptance of the institution of prostitution is based on the notion that men are entitled to sex on demand. It is rarely stated explicitly. As a Nevada politician delicately put it, legal prostitution is "a community service to meet natural needs."[198] "The men need a release, they deserve relief," said a pimp who was a mortician before purchasing a legal brothel.[199] The prevailing attitude of johns is that prostitution in a legal brothel is "not cheating, it's more like going to the grocery store, like stopping at a convenience store," according to one witness.

The same entitlement to sexual access is seen among men in most other countries. As in the United States, men in Brazil, India, and Japan buy sex as long as the activity can be hidden. Cambodian men are assumed to have uncontrollable sexual urges for which prostitution is a safety valve, a catharsis that protects 'good' women.[200] Yet there is no evidence that prostitution protects either women in prostitution or women outside of prostitution from rape. In fact, many psychologists now understand that the catharsis theory is incorrect and that the opposite is closer to the truth: When men buy women in prostitution, they are practicing more extreme sex acts, sexual behavior that they have learned in pornography, which they then perpetrate on their wives and girlfriends. So rather than being a catharsis that is gotten rid of - abusive sex acts are learned and practiced on the bodies and minds of prostituted women, then generalized to nonprostituted women.

Assuming the right to buy any woman, a john in a lap dance club smilingly asked, "You look just like my daughter. Would you like to dance for me?"[201] Women in the brothels report that some johns have pedophile fantasies and pay the women to dress up and to pretend to be little girls, saying "Fuck me, Daddy. I won't tell Mommy."[202] There is an unfortunate belief among the women themselves that pedophile mimicry acts as a catharsis. Unfortunately, it does just the opposite: it reinforces the normalcy of sexual assault of children.

> ...what comes across from men of all classes going to prostitutes is their disregard for the consequences of their actions on the women concerned....[their disregard for the emotional wear and tear of the activity of prostitution for prostitutes.[203]

A woman who prostituted in a massage brothel explained, "Most customers come into a massage parlor thinking nothing is wrong; that it's a job we choose. It doesn't occur to them that we are slaves."[204] Many men describe a nuanced awareness of the physical and emotional suffering of the women that they buy in prostitution. One man said, "There's one overall quality which I seem to notice among all the girls...The girls are what I call *usable*. They're unable to defend themselves and a strong man or woman can take such ridiculous unfair advantage of them."[205]

We found that the prostitution transaction represents and reinforces a massive power imbalance between men and women, one in which johns have the social and economic power to hire women, adolescents, girls or boys to act out their masturbation fantasies.[206] We have begun interviews in several different countries with men who buy women in prostitution. Most of the men to date have confirmed that the relationship in prostitution is one of dominance and subordination.[207] One man told us, *"Prostitution says that women have less value than men."* In prostitution, another man explained,

"She gives up the right to say no" during the time that he has paid for. Another man told us that he clarifies the nature of his relationship to the women he buys, *"I paid for this. You have no rights. You're with me now."* Another explained to us:

> Guys get off on controlling women, they use physical power to control women, really. If you look at it, it's paid rape. You're making them subservient during that time, so you're the dominant person. She has to do what you want.[208]

A bouncer at Las Vegas lap dance clubs angrily described the misogyny of johns: "An exotic entertainer performs dozens of lap dances each night for men who callously grope and utter unwelcome....and degrading statements; men who are disrespectful in every word and action; men who go out of their way to be rude, obnoxious and revolting."[209]

A woman in Seattle described the contradictions in what johns are looking for in women who are in prostitution or pornography:

> Customers liked to think they were getting someone young and fresh. It was weird; they went to a place like Butterscotch's, as dank as it was, and they expected to find their girl-next-door Dulcinea amid the filth. But if we really *were* that innocent, why would we be working in a place where we made money by watching men masturbate to us? The kind of girls they all wanted were daycare teachers, professional cheerleaders, librarians, or nurses – or they were in college studying something midrange and practical-like marketing or communications... They loved to think we were only in Butterscotch's by some kind of weird mistake....[I marveled at the] massive cognitive dissonance necessary for men to believe that the women they'd paid to perform sexual acts were actually *doing* those sexual acts of their own volition instead of for the money. The whole virgin/whore question wasn't about which one they wanted – the secret was that they wanted both, at the same time, in the same girl. That was their 'fantasy girl' – a female they could pay $60 to jerk off to and then take home to Mom.... Every customer wanted to pretend that his cock was the

only one we'd ever seen. But at the same time, they had to know that if we'd take their money, we'd take anyone's.[210]

Describing johns who rate Las Vegas escorts as "hobbyists" who seek a "girlfriend experience," or GFE, Playboy Magazine editors breathlessly promoted women in escort prostitution who can fake the best orgasms and pretend to be in love. In GFE, the woman who is purchased is required to act like a girlfriend, only a more subordinate, obedient girlfriend or wife than the real one, with no commitment or expectations.[211]

It has been established that the sex industry flourishes near military installations.[212] Nevada has more federal land than any other state in the US, including two of the largest military bases, one of which is ten miles from Las Vegas.[213] In southern Nevada, some johns are nuclear test site workers and military personnel. Some are bikers or travelers. In the legal brothels, johns are businessmen, conventioneers, fraternity brothers, truckers, hardhats, oil and mineral drilling crews, migrant agricultural workers, disabled veterans, traveling salesmen, men attending bachelor parties, and politicians.[214]

The population of johns in Las Vegas is likely to be more transient than other cities because of the intense tourism and convention businesses. These men keep track of the openings and closings of various brothels in Nevada, which women are in which brothel, and inform each other of the best bargains in women in Nevada, Costa Rica, and elsewhere on the prostitution tourism circuit.[215] Prices for buying women is a constant topic in Internet chatrooms.[216] Johns share information online about illegal Las Vegas escort prostitution.[217]

Men who are arrested for soliciting women in the Las Vegas streets have the option of attending a diversion program run by the Las Vegas Municipal Court. A few men are over age 50; the majority, according to Howard Meadow, director of the program, are 20-30 years of age.[218]

Johns in the legal brothels also come from other countries. One pimp mentioned that some of his customers were from UK, Canada, and Asia. Often these men are assumed to have more money than regular johns, and so the brothels use Internet advertising to countries out of the US. A significant number of the johns in a legal brothel near Las Vegas were from other countries. Floor managers at another brothel told us that the johns included men from Japan, Germany, and China. A third pimp told us that his legal brothel's biggest spenders were men who located the brothel on the Internet. Johns were from U.K., South America, France, Canada, Vietnam, Korean, and China. They worked in computer technology and mining.

In the northern Nevada legal brothels, one pimp reported that about half of the johns were white European-American, with the other half being Latino, American Indian, African American, Asian, and Middle Eastern.[219] Another pimp reported that 15% of her brothel's johns were immigrants to the US, of which half were from the former Soviet Union and Eastern Europe and the other half were Middle Eastern from Afghanistan, Iran, and Iraq. Many were truckers. A northern Nevada brothel manager told us that most of the johns at his all-Asian women brothel came from California.

In Las Vegas, white European-American men aged 30-50 spend more money in lap dance clubs than other men. These men often like to buy women they deem "exotic" or "something different," but they also tend to be intimidated by men of color. In order to keep these bigoted spenders comfortable, racist policies that exclude African American men and men who speak Spanish are in effect in strip clubs such as Cheetah's.[220]

The trafficking of Chinese, Korean, Japanese, Thai and Filipina women into Nevada prostitution is commonplace. There are large numbers of Asian tourists visiting Las Vegas, including many Asian gamblers. Eighty percent of the highest betting gamblers (those who bet more than $50,000 per hand or dice roll) are Asians.[221] These men are often "comped" women in prostitution to keep them happy while they spend vast amounts of cash at gaming tables in Las Vegas.

One man described his purchase of a young Mexican woman. At first he explained, "I don't want a prostitute who is not having a good time, or who isn't smiling." Then he thought for a minute and remembered a time when he was in a town buying prostitutes on the border between Mexico and the United States with his friends. A young Mexican girl was "dragged" [his word] into the room. Looking apologetically at me, he said, "That was ugly, but we were leaving the next day, so…" And he proceed to tell me about his rape of the girl. Only slightly conflicted about his violently exploitive complicity in the girl's trafficking, at the end of the interview, he looked at me and asked, "Am I a bad guy?"

The man who told me about the foregoing was a white European-American college student. Some johns in Las Vegas are undocumented Latino migrant workers who are employed in the construction or gardening industries. These men, themselves exploited, locate trafficked women who are held in conditions of captivity in the illegal Latino brothels in Las Vegas. This pattern of exploited workers' buying even more exploited trafficked women for sex is seen in other countries. For example single, unaccompanied migrant Cambodian workers in both Cambodia or Thailand find Cambodian women in cheap urban illegal brothels.[222]

Men of all ethnicities have strong inklings and sometimes certainty that the women they rent are under duress or trafficked or in desperate circumstances. According to Maria Jose Barahona who studied men in Madrid who buy women in prostitution, the johns told her that they "go to immigrant women, not trafficked women," re-naming sex trafficking as consenting immigration. There is often a hint that johns are sexually aroused by the desperate circumstances of the women they buy, since so many of them ask, "What's a nice girl like you doing in a job like this?" Johns seem to be seeking evidence of her economic desperation or emotional despair, which then in some convoluted formula enables them to define themselves as eroticized rescuers rather than sex predators. It apparently does not occur to 99% of johns that they could offer her cash without the paid rape of prostitution. His erotic rescue fantasy involves his paying her to perform sex acts which she gratefully surrenders to him, thankful that now she will be able to pay the rent or buy food for her children.

In reality, she is unlikely to be grateful to him for anything. A woman who had prostituted for many years in a Nevada brothel described how she felt about men who buy women in the Nevada brothels: "99% of them fit these words: pig, dog, animal, uncaring, user, slave owner, asshole, mean, thoughtless, rude, crude, blind."[223] Another woman said, "the johns notice the women's bodies but they don't see what is in the women's eyes: "annoyance, indifference, desperation, disdain, agitation… intoxication."[224]

A survivor of prostitution wrote,

It puzzles me why one girl raped by her boyfriend.…has rape crisis centers and counseling and prosecution under the law at her disposal; but if a girl is trafficked, broken, tortured, terrorized and raped over and over, on a daily basis, the many 'customers' who violate her multiple times are not even considered criminals.[225]

5
Trafficking for Legal and Illegal Prostitution in Nevada

A Korean woman who was overwhelmed with credit card debt was led to believe by traffickers that if she traveled to the United States, she could work in the entertainment industry, quickly earn a lot of money and then return home. A college student from a poor family who wanted to impress her new friends, You Mi quickly generated $40,000 in debt. Naively believing traffickers who told her she would pour drinks as a hostess (but would not have to sell sex) for $10,000 a month in Los Angeles Koreatown, she was supplied a fake passport, and once in the U.S. and under the control of traffickers, she was moved between Los Angeles and San Francisco in massage parlors controlled by Korean organized criminals. In 2006 she prostituted 15 hours a day at massage brothels with blacked-out windows and double metal security doors. You-Mi was allowed outside only if escorted by cabbies who were paid by the traffickers. Unable to speak more than a few basic sentences in English, she was unaware of where she was and dependent on her captors for food and shelter. You Mi was isolated, terrorized, and prostituted in a massage brothel under prison-like conditions of debt bondage. After a long struggle, she was finally recognized as a victim of trafficking.[226] Regardless of the nature of the freely made, deceived, tricked, or coerced decision a woman makes to move to another country for prostitution, once she has actually moved or has been trafficked, *she will be* "recruited, transported and controlled by organized crime networks,"[227] Mary Lucille Sullivan wrote about Australian prostitution. These same conditions exist in Nevada.

Andrea, who was born and abandoned in the United States, is another trafficking victim. For 12 years she was a captive of traffickers from organized crime that operated on both sides of the Mexican border, sold mostly to U.S. men. She was transported to Juarez dozens of times. During one visit, when she was about 7 years old, the trafficker took her to the Radisson Casa Grande Hotel, where there was a john waiting in a room. The john was an older American man, and he read Bible passages to her before and after raping her. Andrea described the ceiling patterns of rooms she remembered in other hotels in Mexico: the Howard Johnson in Leon, the Crowne Plaza in Guadalajara.[228]

Because of the ethnic diversity of the United States and because organized criminals are expert at concealing sex trafficking while marketing women as sex products in plain sight, it is often impossible to differentiate between someone who has been harmed in prostitution exclusively within the United States and someone who has been harmed in prostitution in another country, transported, and then again harmed in prostitution once inside the United States.

Prostitution is the destination point for sex trafficking. Legal prostitution is a major contributing factor to the human rights violations of sex trafficking.[229] Where prostitution is legal, states in effect say to the world: we accept the selling of women for sex; we consider pimps and traffickers to be sex entrepreneurs rather than organized criminals; we consider men who buy women for sex to be consumers of sexual services rather than predators. That same message is sent when governments look away from prostitution in their jurisdictions, refusing to enforce existing laws against buying and pimping women.

On the other hand, when a country targets men who buy women in prostitution with serious criminal penalties, as Sweden does, sex trafficking into that country decreases. If you put yourself in the shoes of a pimp, where would you choose to transport women? To a country that has public education campaigns stating that it is a crime to buy sex, as Sweden does, or would you move your "products" or "girls" to a location with legal prostitution, like Nevada or the Netherlands? As trafficking expert Janice Raymond has pointed out, countries with legal prostitution are in effect setting out a legal welcome mat to pimps, johns, and traffickers.[230]

Prostituting women or women not yet prostituted are trafficked to wherever the demand for paid sex exists, to wherever there's a thriving sex industry. Nevada's sex businesses include not only legal prostitution – which is only one tenth of all prostitution in the state - but also escort prostitution, strip clubs, massage parlors, and illegal brothels. Thus almost all prostitution is illegal. Although illegal in Las Vegas, prostitution is so pervasive that writer John Smith described Las Vegas as "the city that has raised sexual exploitation to a billion-dollar legal racket."[231] Under the shelter of the state's aura of legal prostitution and as a result of the tens of millions spent on advertising prostitution, Las Vegas has become a hub of illegal prostitution and sex trafficking that takes place in casinos, hotels, strip clubs, the street, illegal home brothels, and massage parlors.

Driven by men's demand for it, the prostitution and trafficking of women to Las Vegas occurs despite its illegality in Clark County,[232] as a result of zoning decisions biased in favor of sex businesses, a supportive political atmosphere, and the corruption of city county, and state officials and district judges. The huge revenue from illegal prostitution in Las Vegas is the economic driver of the business of sexual exploitation, and it easily corrupts public institutions.[233] Financial contributions "get you juice with a judge - an 'in,'" Ian Christopherson, a lawyer in Las Vegas for 18 years, said. "If you have juice, you get different treatment."[234]

There is an evolving public awareness about the human rights violations of sex trafficking in Nevada and elsewhere in the United States. This awareness and public outrage about trafficking, however, exists primarily for victims who have been transported across an international border. Although physical violence may or may not occur, in all cases of trafficking for prostitution there is psychosocial coercion that happens in contexts of sex and race inequality and under conditions of poverty or financial stress, and often a history of childhood abuse or neglect. Women may legally and seemingly voluntarily migrate from a poorer to a wealthier part of the world, for example with a work permit and the promise of a good job from a "friend" who turns out to be a trafficker. Once she has migrated, away from home and community support, she is dependent on traffickers and their networks. At that point the pimp/trafficker's psychological and physical coercion expands while her options for escape rapidly shrink.

Domestic trafficking - the sale of women in prostitution from poorer to more prosperous sex markets within a single country - can be as devastating for the women as international trafficking.[235] This is true in countries where there is assumed to be significant wealth such as New Zealand and the United States as well as countries where there is more visible poverty such as India and Zambia.

> The apparently civilized transaction between elite prostitutes and their clients in luxury hotels is underpinned by the same logic that underpins the forcible sale of girls in a Bangladeshi brothel. This logic is premised on a value system that grades girls and women - and sometimes men and boys - according to their sexual value.[236]

The economic and social forces that channel young, poor, and ethnically marginalized women into prostitution are evident in post-Katrina New Orleans. Survivors of prostitution and advocates for homeless teenagers in Las Vegas have reported that in the two years following the economic devastation of hurricane Katrina, many young women previously pimped in New Orleans were domestically trafficked to Las Vegas. New Orleans, an economically stressed area with a long history of race discrimination, was the source region for young and poor African American girls. Las Vegas, with its thriving sex businesses, was the domestic traffickers' destination market for the girls.

Wherever there is a market, and wherever they can wrest control from other gangs or from local pimps, organized criminals run prostitution rings both inside countries and across international borders. Traffickers are businessmen who pay close attention to men's demand for prostitution. They obtain the women and girls who supply that demand wherever women are vulnerable because of economic factors and cultural practices that devalue women. The sexism, ethnic prejudice, and poverty that lead to the trafficking of women from Nigeria to Italy[237] - also result in the trafficking of women from one state to another or one city to another in the United States.

Domestic and international trafficking have similar adverse effects on the victim.[238] The psychological harms of trafficking and the resulting traumatic stress are much the same as the psychological harm done to women who are in the legal brothels but who have not crossed an international border. Salgado described a trafficking syndrome resulting from repeated harm and humiliation against a person who is kept isolated and living in prisoner-of-war-like conditions.[239] International trafficking, just like prostitution and domestic trafficking, is extremely likely to result in posttraumatic stress disorder (PTSD). Both domestically and internationally trafficked women experience terror, despair, guilt regarding behaviors that run counter to their cultural or religious beliefs, blame themselves for the abuse perpetrated on them by pimps and buyers, feel a sense of betrayal not only by family and pimps but also by governments that fail to help them.

Social isolation and fears about their immigration status leave women vulnerable to prostitution. If they try to escape violent husbands or pimps they may not know how to access legal or social services. Additional barriers confronted by trafficked immigrant women are absence of services in the language of newcomer groups, discriminatory treatment by law enforcement, and healthcare and social services that are culturally irrelevant or racist.

Most people who had any other survival options would avoid prostitution. We know this because 89% of women in prostitution in 9 countries have told us that if they could, they would prefer to escape prostitution.[240] Similarly, 81% of the women inside Nevada's legal brothels told us that they wanted to escape prostitution.

In spite of women's longing to escape prostitution, some observers have theorized a false distinction between voluntary and coerced prostitution.[241] Pimps bait us with the myth that there is a vast gulf between what they call "freely chosen" prostitution and the physically coerced trafficking of women and children. But is there really such a grand difference, or are some forms of coercion simply more visible than others? How do you know what's behind the mask of a smiling twenty-three year old who is stripping and turning tricks in the VIP lounge of the stripclub? What was her life like before she started prostituting? How many people early on defined her as a little whore while she was sexually abused as a child by family members and neighbors? Did she recently escape a violent husband or partner? Does she have children to support and no job that pays enough? Was she unable to afford to go to college? Was she (for whatever reason) emotionally and economically vulnerable and then tricked and brainwashed by a pimp?

Most johns (those who are not sadists) deny the drastic power imbalance in prostitution and instead maintain that there is choice in her prostitution. Despite evidence to the contrary, and despite common sense, they coyly allege that there is at least a shard of mutual attraction in addition to the cash. The john wants the sordid business of prostitution to appear as if it's "open and informal and bags of fun."[242]

Yet it is the john himself who is the person in the prostitution transaction who has the real choice.[243] He could easily decide not to rent a woman for sex, without sacrificing anything. On the other hand, in almost all cases, women who "choose" to prostitute are doing so out of direct and indirect coercion. Again and again, women say they do not want to prostitute, but they don't see viable alternatives that provide them with an income that even approaches what johns pay them for sex in prostitution.

There is a pyramid-like hierarchy in prostitution. At the top are a very few women who service a few men for a lot of money in a short period of time in their lives - and then they get out, or are bought by one man who supports them. In the middle section of the pyramid are women who need the money, who have had the option of sexual exploitation as a survival mechanism made very real to them by a history of incest or childhood sexual abuse, and who may face an emergency situation such as escaping a violent partner, losing a job, or having children with special needs. The farther you descend in the hierarchy, the greater the numbers of women in prostitution, and the less meaningful any discussion of choice is for them. At the bottom of the hierarchy are the largest number of women in prostitution, those who are the poorest and who have enormously restricted life choices. Many of these women have been physically coerced into prostitution.[244]

The theory that some women choose prostitution ignores the facts that women do not have equal rights with men and that people of color are discriminated against in the United States. Choice depends on the freedom to choose. The lowest earning workers in most cultures are single women who are raising children. In 2005 a minimum wage full-time worker could not afford an average priced one-bedroom apartment anywhere in the United States.[245]

For many reasons, people may "choose" what is deeply harmful to themselves, sometimes because they've grown up seeing themselves in a limited or damaged way. Because they had no alternatives, battered women for many years were assumed to be freely choosing to return to violent partners when in fact they were terrorized into returning under conditions of restricted economic resources. The United States culture itself limits women's ability to reject the consent that is taken for granted.

> The coercion I faced did not involve physical force. Instead, the coercion was emotional and psychological in nature. This included socialization as a sex object, fear of never being accepted in the straight world after being a 'whore,' induction into the sex world 'family' (one that accepts those not accepted elsewhere and a closed system that is difficult to leave), and knowledge that I could not make ends meet and that I would never make the same amount of money (even with a PhD).[246]

A woman in Lusaka, Zambia said, "Yes, I made the choice to prostitute, my children are hungry and I have to feed them." A woman in West Bengal, India said that she prostituted because it was "better pay for what was expected of her in her last job, anyway."[247] Echoing the Indian woman's sentiment, lap dancers in U.S. strip clubs felt the same way about the futility of avoiding workplace sexual harassment. In 2003, Juliana Beasley interviewed and photographed U.S. lap dancers. "Many dancers I interviewed spoke of harassment working in so-called straight jobs," she commented. "Their attitude was, 'If I'm going to be sexually objectified, I might as well get paid for it.'"[248]

Women in U.S. prostitution joke about the "welfare-to-prostitution" trend that has occurred subsequent to the removal of government-assisted educational programs, job training, housing, and childcare. Women are coerced into prostitution by the actions of politicians who remove public supports by shutting down essential social services, who de-fund housing programs and educational opportunities for the poor, and who vote to eliminate food subsidies for vulnerable women and children.[249] Access to these basics in life should not be considered an option or luxury.

Globalized consumerism is relevant to understanding the issues of choice and consent. In many parts of the world, for example Thailand and India, people have a desire for consumer goods like televisions, with social competition to acquire these goods. There is sometimes pressure from families and peers to prostitute because it's the only means of acquiring these items. While young women may not be physically coerced by kidnapping or hunger, nonetheless their naivete is so intense, their options so narrow, that their yielding to prostitution may be tantamount to psychological coercion.[250] The mix of poverty, proximity to financially successful prostitutes who buy their families consumer goods, and a consumer culture – has sometimes dramatically shifted social values regarding prostitution.[251] Jean Enriquez of the Coalition against Trafficking in Women Asia Pacific pointed out that in 2006, global market interests and the militarization needed to protect those interests always seem to accompany shrinking choices and the prostitution/trafficking of women.[252]

Prostitution and trafficking can appear voluntary but are not in reality a free choice made from a range of options. Most discussions of choice and consent erase the fact that prostitution is intrinsically sexually exploitive.[253] Whether or not an individual woman is able to decrease prostitution's physical damage, whether she has relatively more or less money in the bank as backup protection, and whether or not she is protected from sexually transmitted disease by the john's use of a condom - prostitution nonetheless remains harmful. Consent is not a meaningful concept when a woman acquiesces to prostitution out of fear, despair, and a lack of alternatives.[254] Instead of the question, "did she voluntarily consent to prostitution?" the more relevant question would be: "did she have real alternatives to prostitution for survival?" In prostitution, the conditions which make genuine consent possible are absent: physical safety, equal power with customers, and real alternatives.[255]

The pimp's defense is usually that she consented to prostitution. Traffickers offer us the lie that women consent to trafficking. While women may initially consent to prostitution, they rarely know how bad it is going to be, and they never know the prison-like or slave-like circumstances in which they will often be prostituted. A trafficker who operated illegal brothels in 13 states including Nevada was arrested for transporting women from Latin America into prostitution. After his arrest, he argued in court that he gave the women cell phones, let them keep "some of their earnings," and that they had "some periods of freedom."[256] He claimed the women consented but what exactly did they consent to? To being coerced to turn a dozen tricks a day in exchange for having a cell phone, a few dollars, and some pimp-specified amount of freedom?

Pimps themselves certainly don't believe the bogus myth that prostitution is a job choice which is why they find it necessary to employ extreme tactics to deceive, entrap, overpower, and brainwash women. Like military torturers, pimps use forced impossible choices as a way of driving people to hopeless despair. Women are told to choose between harming another person and being beaten up herself. They are pressured to consent to prostitution or their family members will be harmed.

As a woman in a Nevada brothel patiently told me, "no one likes to be sold for sex," whether it's legal or illegal, located indoors or outside, in a gentleman's club or in a gentleman's car. Yet wherever there's prostitution, the pimps' and johns' debate flames up about whether women in prostitution really like it, whether they consent, whether it's voluntary.

An April 2007 antitrafficking bust of an illegal massage brothel in Las Vegas highlights these issues. In the first news report which came out, a law enforcement agent described the women in the brothel: "Some were brought here by force, and some were tricked into coming here with the promise of a good job."[257] Later, a second law enforcement officer stated, "They were all there voluntarily. They expressed they were happy with the money they were making."[258]

The second officer's assessment appears to have been based on statements that were made by a possible trafficking victim who was under extreme duress. Pimps and traffickers carefully brainwash women regarding what to say to police. The women may be threatened with death or deportation if they don't follow pimps' instructions. Prostituted women may have been previously betrayed or even harmed by police. For example, a woman who had been trafficked felt despair when she observed that the police officer who arrested her for prostitution was socializing with her traffickers.[259] It is not reasonable to expect that after being arrested, prostituted women will necessarily be forthcoming. In a culture of subjugation, they can be expected to follow pimps' and traffickers' threats.

Even in cults where there is known to be a lethal tyrant who terrorizes his followers, adult women are assumed to be consenting when they are not. Cult leaders of the Fundamentalist Latter Day Saints polygamous religious group used mind control and terror to enslave women and men. Tragically, as soon as cult members have their eighteenth birthday, even those raised in the cult from birth are assumed by the law or by social services to be voluntarily consenting members. This assumption ignores the fact that many of the FLDS children have never been permitted outside their restricted community - one woman described it as prison-like - and have been instructed to run and hide from strangers. The same tragic assumption that ignores mind control in cults is also made about 18 year-olds who are assumed to be voluntarily prostituting even if they have been under pimp control since, for example, the age of eleven. As an adult woman from the FLDS cult told a reporter, "They don't see you as a victim, it's something you 'chose.' " [260]

It is a challenge for law enforcement personnel to understand the victim's dilemma. Yet the false distinction between voluntary and forced prostitution puts both victims and law enforcement in an impossible bind. According to the terms of the federal Trafficking Victims Protection Act (TVPA), women who are socially marginalized and who have few if any resources or means of protection are expected to bear the burden of bringing legal charges of coercion against violent pimps and traffickers.[261] They're expected to legally challenge organized criminals who are - for good reason - feared by law enforcement and shelter advocates alike. Not surprisingly, since the TVPA went into effect in 2000, very few trafficked women have attempted to prove that they were coerced into prostitution, since it is extremely dangerous for them to do so, and extended federal witness protection is usually not available to them.

Under duress from pimps or traffickers or cult leaders, women hide their coerced status in prostitution just as they do in religious cults. The first lesson taught to women in prostitution by friends who recruit them, by pimps, or by traffickers is to "fake it," or "smile." Prostitution always involves acting. It is disheartening that after years of federally supported antitrafficking workshops to train investigators of trafficking cases, women are still assumed to be voluntarily prostituting because they say to arresting officers that they are happy and are making money.

I talked to people in Nevada, California, Oregon, Arizona, New York, Florida, Canada, the Netherlands, and India about trafficking into Nevada. Law enforcement sources, advocates and service providers all told me that both domestic and international trafficking into Nevada are prevalent and under-recognized. According to a Nevada Brothel Association lobbyist, 90% of women in legal brothel prostitution are from outside of Nevada.[262] Law enforcement sources and survivors alike report that a significant majority of women in the Nevada legal brothels are pimped into them. Together these facts suggest the possibility that a majority of women in the Nevada legal brothels have been domestically trafficked.

Our interviews with 45 women in Nevada's legal brothels were consistent with these reports. 71% of the women had moved from another state to Nevada for the purpose of prostitution, with 58% reporting that they had previously prostituted in other U.S. states.[263] Besides Nevada, they had most frequently also prostituted in California, Utah, Florida, New York, Ohio, Arizona, South Carolina, and New Jersey. Another 20 states were mentioned as well. See Table 4. These interviewees also told us that 50% of the women they knew in the legal brothels had pimps operating outside the brothels.

Pimps from the Nevada legal brothels advertise and recruit women from other U.S. states. Law enforcement agents in Nevada have been unable to use either the Mann Act or interstate prostitution statutes to prosecute pimps when the recruitment and transportation for prostitution occurs to a legal brothel rather than to an illegal brothel.

Table 4. Facts on Transport of 45 Women in Nevada Legal Brothels

Moved to U.S. from another country for purpose of prostitution	14% (6 of 42)
Moved to Nevada from another state for purpose of prostitution	71% (32 of 45)
Prostituted outside U.S.	18% (8 of 45)
Prostituted in other US states Most commonly reported other states* California, Utah, Florida, New York, Ohio, Arizona, South Carolina, New Jersey	58% (27 of 45)
Knows women in legal brothels who are believed to be undocumented immigrants	27% Yes 11% Not sure
Owns a passport	36% (16 of 44)
Uses Internet in prostitution	60% (27 of 45)

*Additional states in the United States where the women were prostituted or from which they were trafficked included Alabama, Colorado, District of Colombia, Georgia, Hawaii, Illinois, Iowa, Kentucky, Michigan, Minnesota, New Mexico, North Carolina, Oregon, Pennsylvania, Tennessee, Texas, Virginia, Washington, West Virginia, Wyoming.

Several confidential sources indicated that organized criminals were involved in a majority of the legal brothels in Nevada. One Nevada brothel owner, for example, controlled several northern California saunas and massage parlors where illegal prostitution occurred. The California massage parlors were a transit trafficking point for women trafficked from Asia, often through Canada, to California, and then to Nevada.

Although the pimps who run the Nevada legal brothels vehemently deny it, women are moved between legal and illegal prostitution in Nevada. A similar legal-to-illegal and back again movement occurs in Australian prostitution.[264] Law enforcement sources in Nevada described the transport of women from financially faltering legal brothels into the lucrative Las Vegas escort prostitution business. Experts are of the opinion that money from illegal escort prostitution in Las Vegas is then laundered through a legal brothel in a nearby county.

Wherever a thriving sex industry exists, children as well as adults are sexually exploited in prostitution. Many young women who prostitute in Las Vegas are minors who are pimped in from the US and Canadian west coasts since johns prefer girls who are young, paying more for them.[265] Children are easier for both pimps and johns to control. Trafficking of 11-17 year old girls to Las Vegas from other states for prostitution is "rampant," according to Las Vegas Metropolitan Police.[266] The children are prostituted in the major Strip hotels and casinos.[267] There has been intense Nevada and Las Vegas law enforcement focus on convictions of pimps and traffickers who prostitute children.[268] U.S. attorneys in California and elsewhere have prosecuted and convicted pimps who move youth women across state lines for prostitution.[269]

It is difficult to accurately assess the ages of younger women in prostitution, whether or not they are in a legal prostitution setting. Young women give their ages as older than they are and it is relatively easy to obtain false identification papers. We heard from several sources that minors were prostituting in legal Nevada brothels. There is likely to be some truth to these as yet unverified accounts since we know that children have been located in both New Zealand and in Australian legal prostitution.

A Portland police officer described a case of a 14-year old Asian child, a recent immigrant to the U.S., who was trafficked to Las Vegas by a pimp who moved her and other girls on a circuit that included Portland, Seattle and Las Vegas. Las Vegas police brought charges of kidnapping, pandering, living off the earnings of a prostitute, and furnishing transportation against the pimp. The Portland police also charged him with rape and sodomy.

Pimps use verbally and physically coercive techniques to entice, brainwash, and ultimately control women in prostitution. A fairly common pimp maneuver is to pose as an aspiring rap star in order to impress naïve young women. All he needs, he tells her once she is persuaded that he loves her, is for her to earn just enough by prostituting to pay for his recording demo. Soon after, she is 'seasoned' into prostitution by gang rape, brutal physical violence, and toxic verbal abuse. This was the likely scenario of Lee Laursen, a young woman from Utah who met aspiring rapper/pimp Jason Mathis. At first she went with him willingly, but at another point in time, perhaps when she realized the great danger he posed to her life, she phoned her family from Las Vegas saying that she was being held by Mathis and other men against her will. Police located her, informing her that Mathis was dangerous and that he had been charged with two homicides. They tried to convince her to return to Utah but she seemed brainwashed, they said, and refused to leave Mathis. Four months later she was murdered. After her death, her cell phone was located, and police fielded calls from johns asking for 'Alana,' the name Laursen used to advertise escort prostitution on the Internet.[270]

Young African American women may be attracted to the hip hop scene with dreams of becoming stars. They are often coerced to perform acts on camera that are far beyond their comfort zones, and they are prostituted by the misogynist men who are both hip hop stars and pimps. Karrine Steffans has written an account of her life inside the hip hop industry which included her victimization by domestic violence, time spent in Ice-T's 'pimp room,' sexual assaults, and strip club prostitution.[271]

In a case of domestic trafficking, eight pimps were charged in 2005 with conspiring to transport minors for prostitution in Las Vegas, via a network that moved children and adults for rent as prostitutes in Massachusetts, Florida, New York, New Jersey and Nevada. The case included money laundering.[272]

In 2006, the FBI, the Department of Justice and the National Center for Missing & Exploited Children reported the conviction of a Reno pimp who trafficked and prostituted 14 and 16-year old girls in both California and Nevada.[273]

A woman reported that she was trafficked from a southern state to a number of different brothels in Nevada. When she resisted, drugs were used to sedate her. During those times that she was drugged, she was mentally programmed by pimps to participate passively in prostitution.[274]

A woman reported being kidnapped outside a San Diego club, not having realized that an elaborate plan had been set up by her new acquaintances to abduct her as she left the club. She was conditioned, brainwashed, terrorized, and then domestically trafficked to Las Vegas. She suffered a psychological breakdown that was perpetrated by pimps' brainwashing as the crucial element in her trafficking. She was isolated, sensorily deprived (she was kept in a dark room, blindfolded, with no human contact), subjected to gang rapes, her identity changed, and she was physically debilitated and terrorized. She was brainwashed through the repetition of relentless messages about her worthlessness except as a saleable sexualized object. She was driven past her younger brother's school with a gun to her head while her traffickers pointed out how easy it would be to kill her entire family. This trafficking survivor was profoundly disturbed by observing a second young woman who was forced to beat up a third woman. She learned that she could trust no one. She sensed that the politicians and public figures who bought her for sex knew that she was there against her will. Although she made tentative attempts to ask them for help, the johns who used her never spoke a word to her.[275]

In another case, a pimp was arrested for money laundering after he made real estate purchases with proceeds obtained via prostitution and from identity theft.[276] Another man laundered his prostitution monies into an account held by his lawyer after trafficking women from Portland to prostitute in Las Vegas.[277]

6

Illegal Escort and Strip Club Prostitution in Las Vegas

"If I hadn't had the experience I had working in vice," a Las Vegas police officer said, "I would probably think that strip clubs, escort services were cool... But I think the layperson has no idea what goes on in strip clubs and escort services. I've seen the boyfriends, the pimps. I see how the women are treated. It's given me a whole different perspective."[278]

The economic success of the sex industry depends on the public's complicity in denying the harm of prostitution or at least their naïve acceptance of blatant lies. The myth that 'prostitution is a victimless crime' or that prostitution isn't happening at all – are beliefs that the public uses to avoid knowing about the very disturbing realities of prostitution. Some assume that it must be legal since it is advertised as 'entertainment' and 'massage' in 173 pages of the Las Vegas phone book.

Strip clubs are an integral part of prostitution in Las Vegas yet the fiction is maintained that strip clubs do not include acts of prostitution. The lines between prostitution and sexually exploitive activities such as stripping have become blurred. The amount of physical contact between customers and women who strip has steadily escalated in the past 15 years, along with an increase in sexual harassment and physical assault. Sexual harassment, abuse, and assault are intrinsic to stripping and strip club prostitution.

Touching, grabbing, pinching, and fingering of dancers by johns removes any boundaries that historically existed between dancing, stripping, and prostitution.[279] Kelly Holsopple documented the verbal, physical, and sexual abuse experienced by women in strip club prostitution, which included being grabbed on the breasts, buttocks, and genitals, as well as being kicked, bitten, slapped, spit on, and penetrated vaginally and anally during lap dancing.[280]

In most strip clubs, customers can buy a lap dance where the dancer sits on the customer's lap while she wears few or no clothes and grinds her genitals against his. Although he is clothed, he usually wears a condom and expects ejaculation. The lap dance may take place on the main floor of the club or in a private room. The more private the sexual performance, the more it costs, and the more likely that violent sexual harassment or rape will occur. At one strip club, a woman explained, "We know when [prostitution] happens [during private lap dances]. Then four songs are played instead of two."[281]

There are a range of prostitution activities that occur in strip clubs, most frequently body-to-penis friction, masturbation, and oral sex. Johns occasionally wear pants with no underwear so that their genitals can easily be exposed for masturbation or oral sex. They also masturbate themselves if the dancer does not.[282]

Some johns who are strip club regulars pretend that they are involved in a serious, long term relationship with the stripper, bringing her expensive gifts.[283] The danger in this self-deception is that it often results in johns' stalking of women who work at strip clubs, an ever-present risk for the women. Johns who are described as 'Relaters' are dreaded by women in prostitution because they demand more from her. The 'Relater' needs to feel that he is actually making a human connection with the woman he's renting, that she really likes him and finds him attractive. Sarah Katherine Lewis explained,

> Men who use prostitutes and say things like *I made sure she came, too,* and *I only see girls who actually enjoy their work,* are Relaters. They are loathsome, and most working girls avoid them as much as they can afford to, because Relaters need to think they aren't, essentially, *customers.* They like to pretend they honestly respect the women who get them off for money. But their insistence on obtaining personal information from their providers is telling. They just want *more* intimacy for the same price, however they can wrangle it....*Shut up and let me do my job, cow,* I mentally commanded him. *Stop trying to relate to me on a human level. We both know that's impossible. You're a cow and I'm a milkmaid.*[284]

Prostitution occurs in strip clubs in Las Vegas, often in the private dance rooms, and sometimes in the public lap dance rooms, where up to 20 glazed-eyed men sit together in a semen-stained room guarded by bouncers, while women grind the men's laps. There are occasional prostitution arrests in Las Vegas strip clubs, as in the case of the Olympic Garden.[285] Johns exchange information online about strip club prostitution in Las Vegas.[286] At the Crazy Horse Too strip club, one john reported that prostitution takes place in the Emperor's Room, one of the strip club's private rooms. The Crazy Horse Too has been connected to organized crime in New York and in Illinois.[287]

A young woman described her metamorphosis from the perspective of "Oh yeah, I'm going to be a stripper and be cute and dance on stage" to "hey I can escort and it's no big deal." "At first," she said, "They all tell you, 'In just two weeks you can walk out with $15,000.' All strip clubs in Vegas have lap dances now, it's extremely raunchy - anything goes." New dancers are groomed for prostitution by strip club pimps, who teach the women how to sexualize their interactions with men in order to maximize their income. Club owner pimps watch the women and waive stage fees for those who are big earners.

Strip club dancers who do not permit direct sexual contact or prostitution do not earn much income. In any club where lap dances or private dances are available,[288] a dancer who simply performs on the stage might end up in debt. Strip clubs charge a daily 'stage fee' (averaging $200 in 2006) for stripping. In addition to that fee, dancers are required to tip disc jockeys, bartenders, bouncers and managers at the end of their shifts. A woman who stripped at Cheetah's in Las Vegas earned no money because she did not permit any sexual contact. When she complained about this, a brothel employee told her, "Well, you need to grab these guys by the crotch to get their attention."[289]

Lily Burana worked in Cheetah's strip club in Las Vegas where "…everybody's trickin.' There are 2 sets of rules at Cheetah's as everyplace – the ones they tell you and those that you actually work by." [290] She noted that the business of stripping is about "skillful manipulation, and if everybody were in on the scam, stripping would quickly become obsolete."[291]

A strip club bouncer described the strip club scene in similar terms, "The entire industry is illusion and fantasy. Who would want to hear about how fat, pathetic, and smelly [the strip club johns] are?"[292]

As in other types of prostitution, men who buy sex in strip clubs are often paying for women to pretend to like them. Ironically, this GFE - girlfriend experience - prostitution is considered by the women to be more stressful than a 5-minute blowjob, because of the emotional pressure from the john to act differently from how she feels. The quick blowjob doesn't require as much acting, and it's over with quickly.

Strip club managers pimp women in a network controlled by organized crime that stretches across the United States. Both Las Vegas and Portland police departments report that many of the Las Vegas strip club lap dancers are pimped on a domestic trafficking circuit between Portland, Las Vegas and other U.S. and Canadian cities. These circuits are controlled by strip club owners who are most often pimps and who are frequently affiliated with organized crime groups. Illustrating strip club pimps' connections with legal prostitution, a Nye County legal brothel was recently purchased by two men who were owners of several strip clubs in New Orleans.[293] The six legal Nevada pimps I interviewed were quite reluctant to talk about their previous employment. Several hinted at connections with organized crime.

5.6 million convention goers visited Las Vegas in 2003, bringing approximately $6.5 billion to the city.[294] Many of these visitors were men without their wives or partners. A legal brothel owner stated in 2003, "Our business is a pittance compared to prostitution in Las Vegas. I believe there is $500 million in [prostitution] business done illegally in Las Vegas every year."[295] His estimate was low. Three law enforcement sources, two investigative reporters, and two cab drivers who received extortion monies- separately estimated that the sex industry and its ancillary operations in Las Vegas generate between $1 and $6 billion per year. These estimates include both legal and illegal activities such as legal lapdancing, extortion monies paid to taxi drivers for delivery of customers to specific strip clubs, illegal escort and massage prostitution, and tips to valets and bartenders for procuring women. In approximately 1990, a valet job in Las Vegas was so lucrative that it cost approximately $250,000 to buy access to the job.[296]

Second only to gaming, strip clubs are major sources of revenue in Las Vegas. Jack Sheehan analyzed the income from private booths in strip clubs where prostitution happens. In 2004, he found that these booths were estimated to generate $500 per hour. If 15,000 women in Las Vegas have sheriff's work cards permitting them to dance topless or totally nude, and if a third of those women are in the strip clubs on a given day and if they bring in 2 hours' of private room money ($1000), then the money generated from this activity is $5 million a day or $1 billion, eight hundred thousand a year.[297]

There are approximately 40 strip clubs in Clark County.[298] Each employs approximately 100 women daily who average $350 per day in tips. That sum alone is $430 million per year.[299]

Cabbies report that at least 25% of their customers ask to "get a girl" or "find some action."[300] If a cabbie does not procure, his income averages $35,000 a year. If he procures and pimps, he can easily earn at least $150,000 for pimping women to johns, delivering customers to strip clubs, and/or selling drugs out of his cab.[301]

Cheetah's is an average-sized strip club in Las Vegas. A bouncer employed at Cheetah's calculated the business's 2003 income. According to his calculations there are approximately 1800 customers on a Friday evening (9pm-5am), each paying a $20 cover charge. Cover charge income on that shift was $36,000. Drink sales average $25,000 per shift.[302] A Friday's income at Cheetah's is likely to be $72,375. Assuming that other days of the week are likely to generate less income, perhaps $20,000 less, Cheetah's gross annual income (it's open most days of the year) is $18 million.[303]

The cash generated in strip clubs - both legally and illegally - attracts organized criminals. A dizzying web of sexual exploitation against dancers, organized crime, political corruption, and bouncer violence against johns, was evident at the Crazy Horse Too strip club in Las Vegas. As writer and former Las Vegas City Councilman Steve Miller has written,

The Crazy Horse Too has been the repeated scene of violence, prostitution, extortion, and coercion. Frederick Rizzolo, the owner of the Crazy Horse Too was convicted of a felony for beating a patron with a baseball bat in 1985. Records indicate 737 police responses in 3 years but no criminal prosecutions."[304]

Las Vegas Mayor Oscar Goodman accepted $40,000 in campaign contributions from Frederick Rizzolo. District Attorney David Roger accepted and then returned $13,000 in campaign contributions from Frederick Rizzolo, and subsequently dropped five cases against Crazy Horse employees brought by Las Vegas Metro Police.[305] One of the defendants in the Crazy Horse investigation was Rizzolo's associate Vincent Faraci, a reputed Bonanno crime family soldier and the son of New York mobster "Johnny Green" Faraci. [306]

The same pattern of organized criminal violence is evident in Australian strip club prostitution. The owner of a global strip club chain Spearmint Rhino was known to be connected to La Cosa Nostra in New York, with felony convictions in the United States, and licenses for prostitution revoked in the UK.[307] A strip club license investigator in Melbourne, Australia was violently assaulted by owners and managers of a strip club.[308]

Adverse effects of stripping on women who strip

In Nevada as in Australian prostitution, women are transported between legal and illegal prostitution. Sheri's brothel in Nye County where prostitution is legal, offers free passes to Sheri's Cabaret, a nude strip club in Clark County where prostitution is illegal. When business at Sheri's legal brothel was slow, some women from the brothel danced and prostituted in the Las Vegas strip clubs. Others moved in the other direction: from strip club to legal brothel. A woman was forced by her boyfriend, who used verbal and physical coercion, to first strip in a club, and eventually to prostitute in a legal brothel. Under the age of 18, in the brothel she was coerced to prostitute in shifts that lasted from 12-14 hours at a time. She was afraid not only of her pimp/boyfriend but also the legal pimp who ran the brothel. She also feared the johns. She felt ashamed, and she was emotionally numb.[309]

The sexual and physical exploitation of stripping include objectification, verbal abuse, loss of dignity, and humiliation. A man who worked at a strip club for many years reported to the author that most - as many as 80% - of the young women who came to work in the Las Vegas strip clubs could not survive a week because they could not tolerate johns' verbal and psychological abuse.

Pimps and customers both perpetrate domestic violence against women who are dancing or prostituting in strip clubs. One woman described beer bottles being thrown at her while she was onstage. Las Vegas Metro police report frequent physical assault charges filed by women who list their occupation as "dancer."

Customers also experienced violence in strip clubs. Bouncers who used steroids and methamphetamines regularly assaulted customers, sometimes causing serious injury. Murders have been reported at stripclubs.[310]

Women's level of poverty is one factor determining how long they remain in strip clubs.[311] Some women begin stripping as young as age 13. Others are undocumented immigrants, runaways, and women in prostitution who need a break from the streets.[312]

Stripping can have unanticipated negative effects on the women's relationships.[313] Many women in lapdance or strip clubs, like women in prostitution, have difficulty enjoying sex with their non-john chosen partners.[314] Disconnection and often dissociation is required in order to perform stripping or acts of prostitution in clubs.[315] Women describe the emotional effects of stripping with the same words they use to describe the effects of prostitution: exhaustion, rage at men, permanent shell-shock, inability to calm down, and an intense self-hatred.[316]

All-nude strip clubs/lap dance clubs are common in Las Vegas. Performing the gynecological demonstration for the men in the front row of the all-nude strip clubs in Las Vegas requires a somatic dissociation.

Spreading the labia to permit men to look inside women's vaginas sends the message that "the stripper has no private self, that everything about her is open to inspection and invasion, that her very soul is up for grabs, that she can be turned inside out, and that she has no boundaries, no conditions, and no limits on who she lets in."[317]

A woman who worked as a lap dancer at Cheetah's told Brent Jordan:

> I can no longer tolerate the touch of a man, any man. A man's touch has come to represent labor and degradation, and a sad, sick feeling of desperation and despair. Every sort of hateful, spiteful, rude, venomous remark, I have endured. Vile anger, vomited from the crude, the resentful, the desperate and desolate, has been heaped upon me until I have choked on it. I have come away with, not hate, but worse, a numb disinterest.[318]

Las Vegas police priorities do not generally address strip club prostitution. Las Vegas Metro Police Captain Anthony Stavros set out a specific list of priorities: juvenile prostitutes and their pimps, and street prostitutes and their pimps. "Between those four, we stay pretty busy."[319] As another law enforcement source told me, "Compared to the arrests of women for solicitation, we arrest very few johns. And almost no johns at all in casinos or hotels."

In Las Vegas, police focus their attention on street prostitution and what they describe as 'juvenile prostitution' but which would more accurately be described as sexual assault of a child. While police routinely charge pimps with living off the earnings of a prostitute, providing transportation to a prostitute and living with a prostitute, all felonies, indoor prostitution is for the most part ignored: prostitution that involves johns renting adult women in strip clubs and hotels. Pimps are usually convicted of only one or two of the charges and too often they end up back on the streets within a few months, or at most a few years.[320]

7

Domestic and International Trafficking for Prostitution and Organized Crime in Nevada

International trafficking to and from Las Vegas and other locations in Nevada is an increasing concern of local and federal law enforcement, the Department of Labor, and social service agencies.[321] There is evidence of organized crime in strip club prostitution, escort prostitution, legal brothel prostitution, pornography production in brothels and clubs, and in the domestic and international trafficking of women for the purpose of prostituting in a range of locations throughout Nevada. [322]

Despite these facts, there is a persistent myth that legalizing prostitution decreases organized criminals' involvement in prostitution. Nothing could be farther from the truth. Sex trafficking cannot happen without the involvement of organized crime.

Organized criminals in Nevada's sex industry are frequently internationally networked, both fragmented and at the same time collaborative, constantly adapting to maximize profits. With potential high payoffs and relatively low risks, the sex trafficking industry attracts many different types of criminals, many who regularly engage in criminal activity, others whose criminal activity is sporadic, and some for whom criminal activity is new or infrequent.[323]

This phenomenon is not unique to Nevada or to the United States. Organized criminals laundered money in Turkey after obtaining it by trafficking women in Dutch legal prostitution.[324] An Australian Crime Commission found that serious and organized crime was well entrenched in regulated prostitution, that Australia was "very attractive to human traffickers," because organized criminals searched globally for locations where profits are the highest and risks the lowest.[325] Clearly, legal prostitution decreases pimps' risk of arrest for pimping and johns' arrest for soliciting prostitution.

Wherever prostitution is legal, sex trafficking from other countries is significantly increased into both legal and illegal sex businesses in the region. Women are trafficked from other countries into Nevada's legal brothels via quasi-legitimate 30-day work or tourist visas. These are the same methods used to traffic women into Australia and the Netherlands.

Prostitution is legal in the Netherlands yet organized criminal involvement in both the legal and illegal sex industry has drastically increased since Dutch brothels were declared legal by establishing prostitution zones. The Dutch government has unsuccessfully struggled to control the influx of organized crime to areas where legal prostitution flourishes.

In 2004, Amsterdam's mayor acknowledged that the city's legal prostitution zone had become a magnet for traffickers and that the prostitution zone was unsafe for women. He admitted that the prostitution zone had become a "haven for traffickers and drug dealers, and had not achieved its aim, to break the links between prostitution and organised crime."[326]

According to two sources outside the Netherlands, between 70% and 80% of all women in Dutch brothels were trafficked from Eastern European, African, and southeast Asian countries with devastated economies.[327] In 2004, observers noted that many Nigerian girls were being trafficked into the Netherlands, often to its legal brothels.[328]

The alarm about international crime organizations that have infiltrated the Netherlands, the increase in trafficking of women and children into the Netherlands, and distress about the Netherlands' image as an international sex tourism destination has caused some to question whether legal prostitution was ever a good idea in the first place. Frank de Wolf, a Labor Party member of the Amsterdam City Council who is also an HIV researcher stated, "In the past, we looked at legal prostitution as a women's liberation issue; now it's looked at as exploitation of women and should be stopped."[329]

Trafficking has also increased into Australian prostitution where criminals transport women from Thailand, Korea, and China, among others into Australian legal and illegal brothels.[330] Like Nevada, the Netherlands, and Australia - Germany also has legal brothels. Turkish traffickers in 2007 moved women from Germany, Poland, the Netherlands, Romania, Bulgaria, and Ireland into both German legal apartment/brothel prostitution as well as into Dutch legal brothels.[331]

In Nevada, 27% of our 45 interviewees in the Nevada legal brothels believed that there were undocumented immigrants in the legal brothels. Another 11% said they were uncertain, thus as many as 38% of the women we interviewed may have known of internationally trafficked women in Nevada legal brothel prostitution. One woman hinted to me that she had herself been trafficked from China. She had entered the United States in northern California, where a trafficker controlled a number of saunas and health spas that were cover operations for prostitution/trafficking. After prostituting for a period of time in an illegal California health spa, she was moved to a legal Nevada brothel.

Why would the Nevada legal brothels run the risk of using women in prostitution who had been trafficked from another country? It may be because the legal brothels have had difficulty recruiting women. The numbers of women willing to prostitute in the legal brothels has decreased in recent years, partly due to the danger of HIV, to an increased awareness of the harms of prostitution, to the greatly increased illegal sector, and partly due to the prison-like atmosphere and financial exploitation in the legal brothels.

Kuo suggested that this difficulty of locating women to prostitute in the legal brothels is likely to increase trafficking from other countries.[332] Many women who prostitute prefer illegal to legal prostitution. Since most women are pimped into the legal brothels by husbands, boyfriends or pimps, some prefer to avoid the double whammy of pimping that they are subject to in the legal brothels where they are subordinate to both legal and illegal pimps.

Legal prostitution combined with a widespread tolerance of illegal prostitution "means that Australia is now recognized as a destination country by traffickers, international governments, and buyers alike."[333] This also appears to be true for Nevada. We saw for ourselves and also heard accounts of internationally trafficked women in the Nevada legal brothels.

A pimp I interviewed stated that a Romanian trafficker inquired about purchasing a legal brothel in Nye County. Another pimp in northern Nevada told me that he had an offer to purchase his brothel from Russian organized crime. This is a cash business, after all, he pointedly explained to me. Several informants told me that they knew Russian and Romanian women who were trafficked into a Nye county brothel. I spoke briefly with a very nervous woman from the Czech Republic who lived at a southern Nevada brothel.

We visited a legal Nevada brothel where there were approximately 30 Chinese women who did not speak English. An assistant pimp told us that the women entered the United States on 30-day work permits. The permits frequently expired while they were prostituting. Several women in a nearby brothel told us that most of the women in that particular brothel had been trafficked from southern China to illegal massage brothels in and near San Francisco, and from there to the Nevada legal brothel.

According to these witnesses, the Chinese women were threatened with being thrown out of the brothel, reported to immigration officials and deportation if they did not agree to service the high numbers of buyers they were pimped to and if they did not accept whatever paltry amounts of money the pimp chose to pay them. The women were coerced into offering johns oral sex without condoms. They were also required to undercut prices of nearby brothels. We reported our observations about this brothel to the FBI in January 2006.

Internationally trafficked women are also transported to the more controlled, sometimes captive, indoor environments of both legal and illegal brothels and massage parlors. In 2005, two Thai women were arrested with fake identity cards in Las Vegas for soliciting prostitution in a massage parlor. It appeared likely that they were trafficked.

Mary Lucille Sullivan has discussed Australia's difficulty in regulating illegal prostitution, noting that criminals traffic women and girls between the two sectors - legal and illegal.[334] We observed the same phenomenon in Nevada where the boundaries between legal and illegal prostitution are blurred. Las Vegas is a global destination for prostitution tourism, also called sex tourism.[335] Wherever there are large numbers of tourists and visitors arriving in a destination to buy women for sex, it is highly likely that sex trafficking also occurs. Pimps move women to wherever market demand for paid sex exists, ignoring prostitution's legal or illegal status.

Organized criminals operate independently but in cooperation with other individuals such as government officials, pimps and even other mafia groups."[336] Organized criminal groups are often linked, and can thereby facilitate international trafficking. They are skilled at operating outside the law and evading prosecution. In Australia, there are clear connections between organized crime and legal prostitution. For example, a majority of the shares in a publicly traded Australian brothel were owned by John Trimble who is a pimp with established connections to the Australian Mafia, and who is also a nephew of a prominent Australian Mafia figure.[337]

Strip clubs and escort services in Las Vegas are "a natural cash business for the mob," Clark County Sheriff Young told a reporter.[338] The interstate network of strip club owners who pimp women on a U.S. circuit is one sector of Nevada's organized crime. Women are moved from state to state in a stripclub circuit because johns seek "something new," and because pimps move women out of state subsequent to arrest in a particular location. Several Las Vegas club owners are involved in this kind of network.

There is a constant transfer of funds from organized criminals in strip clubs to politicians' campaign funds. This occurs not only in Las Vegas but across the United States. For example, two men in the Clacurcio family who owned several Seattle strip clubs were charged with illegal campaign contributions to city council members. At the time this was written, local and federal law enforcement were investigating prostitution and unsolved homicides at the family's strip clubs.[339]

We learned of a number of instances in which money was laundered by a pimp who owned both legal and illegal brothels in southern Nevada. He transported women in the legal brothel into illegal escort prostitution, because when large conventions were held in Las Vegas, more money could be made from illegal escort prostitution than from legal brothel prostitution. The proceeds from escort prostitution were then reported as legal brothel income.

A similar laundering of cash from illegal prostitution occurs in Australia where, like Nevada, prostitution is both legal in some locations and illegal in others.[340] Widespread tax evasion from not only the legal brothels but from strip club and other prostitution was reported in 2007. In a noteworthy understatement, an Australian tax-office spokeswoman said, "Many in the adult industries have poor record-keeping practices (and) there are low levels of compliance among escorts, adult entertainers and brothels."[341]

Organized crime has noticeably decreased in Sweden since the 1999 law on prostitution went into effect. Aiming at the abolition of prostitution, Sweden vigorously prosecutes men who buy women in prostitution as well as pimps and traffickers. Therefore Sweden is not a welcoming environment to organized criminals.[342]

International gangs engage in a wide range of criminal activities in addition to prostitution and trafficking. A number of law enforcement agencies have reported the presence of Israeli organized crime in illegal Nevada prostitution.[343] Israeli and Eastern European organized criminals operated a Palm Springs escort prostitution business that pimped 240 women throughout California, Nevada, Oregon, and Arizona. In addition to prostitution/trafficking, Israeli crime syndicates are involved in loan sharking, extortion, money laundering, and illegal gaming in Las Vegas.

Women in prostitution are used by Israeli criminals not only in prostitution but in other illegal activities. In one case, a woman was transporting a large amount cash from Las Vegas to a high-ranking member of an Israeli organized crime syndicate in Spain. The "Jerusalem Network" crime family, who were suspected of conspiracy, extortion, and money laundering in Las Vegas, also laundered money via lawyers in Miami and extorted from car dealers in Beverly Hills. [344]

Outcall or escort prostitution in Las Vegas generates tens of millions of dollars annually, according to several law enforcement sources. Many outcall/escort businesses are operated by Russian, Armenian and Mexican organized criminals. In recent years, the Gambino family of La Cosa Nostra has attempted to re-enter escort prostitution in Las Vegas, but for the most part, international organized crime has replaced the older crime families in Nevada's sex businesses.[345]

Eastern European organized criminals have been involved in Las Vegas escort prostitution for a number of years.[346] A 2005 case involved traffickers who ran Elite Escorts, through which women were internationally trafficked into Las Vegas. Using 80 phone lines, traffickers used the Internet to conduct most of their business and were suspected of pimping, pandering, perjury, loan fraud, money laundering, falsifying income tax returns and grand theft. Prostitution monies were laundered into real estate investments. The escort prostitution gang trafficked women to 22 US cities before its owners, Elena Trochtchenkova and Rady Abdel Salem Abbassy, were arrested on federal interstate prostitution, money laundering, and tax evasion charges.[347]

According to writer John Smith, mobbed-up gangsters are at the front doors of smaller clubs in Las Vegas, but are in the back offices of the larger clubs. Smith told me that the presence of organized criminals is more blatant in smaller strip clubs because in the larger operations there is much more at stake so the criminal organization is better concealed.

Mexican and Russian pimps/traffickers manage and control swingers' clubs in Las Vegas which in many cases are brothels with trafficked women prostituting in back rooms. A Russian woman trafficked her niece from Russia to Los Angeles and from there to Las Vegas for prostitution.[348] Las Vegas Police have located illegal brothels in Las Vegas apartment complexes with trafficked Eastern European women.[349]

Korean and other Asian women are trafficked to the United States via Canada. Vancouver Detective Constable Raymond Payette stated that Chinese and Korean organized criminals trafficked women into Vancouver, and from there, they transported both native Canadians and women from

Asia on a circuit from Vancouver to Portland to Las Vegas or Vancouver to Seattle to Las Vegas for the purpose of prostitution.

While suspected traffickers smuggled prostituted Korean women into California massage brothels, johns post specific information on the Internet about illegal Korean massage brothels in Las Vegas.[350] After their entry into the United States, traffickers moved women from San Francisco to massage parlors in Las Vegas, Dallas, New York, and Boston.[351] They were arrested in 2005.

In 2006 the Las Vegas Rape Crisis Center told us that they had provided assistance to sex trafficking victims from Russia, Haiti, Korea, and Mexico. Las Vegas Metropolitan police reported cases of trafficking of women from China, Korea, and Brazil. A Las Vegas legal clinic reported that they had encountered Thai, Filipina and Chinese women who were victims of trafficking.

Asian organized criminals have been arrested for trafficking Chinese women into California and Nevada for prostitution.[352] According to Las Vegas Vice Detective Aaron Stanton, most of the illegal massage parlors in Las Vegas provide johns with internationally trafficked Asian women. The women usually speak only enough English to negotiate prices for masturbating or performing oral sex on the men. Detective Stanton stated that these are difficult cases because of the women's mistrust of police and mistrust of English speakers in general. The women are also terrorized by traffickers' death threats against themselves and also against family members at home. In order to bring any charges against a trafficker, victims need physical protection and emotional support, including an advocate who speaks the victim's first language, and law enforcement agents who understand the brainwashing that is perpetrated against victims of trafficking.[353]

Asian gangs are the largest organized crime group in Las Vegas, controlling massage parlors and private illegal brothels in suburban homes. The crime groups running the Asian massage parlors have highly organized operations guarded by armed security personnel who are employed by traffickers.

Massage parlor traffickers are accustomed to police busts and have attorneys in their employ who are knowledgeable about antitrafficking legislation. Traffickers' colleagues have been known to attend federally-sponsored antitrafficking workshops, posing as advocates. Familiar with antitrafficking laws, the traffickers manipulate the system to ensure that arrested women will be bailed out and promptly returned to prostitution.

Asian traffickers - regardless of their own affiliation with Chinese, Korean, Japanese, Filipino, Vietnamese, Laotian, Malaysian, or other Asian organized crime groups – traffic women from many Asian countries into U.S. prostitution. Chinese syndicates may partner with the Japanese Yakuza in a globalized industry that exploits the racism of Asian prostitution and where individual women's ethnic heritages are deliberately blurred. Women are advertised as "Chinese Geisha girls" or "Thai/Filipino." In Hawaii for example, an escort agency advertised "Chinese-Thai-Filipino Spoken," suggesting the existence of linked criminal networks.[354]

The Yellow Pages advertising Asian women in prostitution, in thinly-veiled ads for "entertainment" and "massage," is racist as well as sexist. Asian women are sexually objectified in advertising for prostitution. But at the same time the Yellow Pages are an indicator of the source countries of women who may have been trafficked to Las Vegas. The most frequently advertised Asian women for sale in Las Vegas were Thai, Chinese, Japanese, and Korean – the same groups of women who are trafficked into Las Vegas. Other Asian women advertised for prostitution were Cambodian, Eastern Asian, Filipino, Hong Kong, Indian, Malaysian, Singaporean, Taiwanese, Tokyo, and Vietnamese.[355]

Las Vegas has a growing Latino community which comprises approximately 34% of the population of Clark County. Many Latinos are immigrants from Mexico, El Salvador, Ecuador, Honduras, and other Latin American countries. A majority of Nevada's Latinos are employed in the construction and agricultural sectors of the economy. A minority of Nevada's Latino population are people who are undocumented. Of that undocumented group, many are trafficked women. Some Latinas are trafficked and are engaged in both domestic servitude and prostitution, in that prostitution is expected of them as part of their role as servants. Violent gangs from El Salvador exploit the market created by undocumented male workers' demand for prostitution in the Las Vegas region.

Several Nevada cases of coerced agricultural labor and domestic servitude have surfaced. When labor trafficking cases involving women victims are carefully investigated, the women are usually sexually exploited as part of the activities for which they are trafficked. According to a law enforcement source, an undocumented man who worked in the construction industry brought his wife to Las Vegas from northern Mexico, then pimped her to five of his coworkers. He and they viewed her as his property. Women who are trafficked as maids are often expected to tolerate rape by employer-traffickers in addition to performing other household tasks. According to a second law enforcement source, when women are bought and used as domestic servants, "all are sexually assaulted within the first month." In these cases, women were trafficked into conditions of labor servitude that included prostitution-like activities.

Latinas are also trafficked into illegal brothel prostitution. Las Vegas police raided an illegal brothel with Latinas who were trafficked from several different Latin American countries. The brothel admitted only Spanish-speaking johns. A pimp who was arrested at the brothel was registered in Nevada as a California Tier 2 sex offender.[356] It is likely that this Latino brothel was owned and controlled by organized criminals from Mexico.

The conditions in these Las Vegas brothels are like the closed brothels of Cambodia, where girls and young women are kept under violent coercion and captivity. There is greater poverty in Cambodia than in Nevada, which increases the vulnerability of women in Cambodian prostitution. However, the Japanese, Chinese, and Cambodian men who buy girls in Cambodian closed brothels don't hide behind the hypocrisy that they are buying consenting adults.[357]

The men who go to the Las Vegas closed brothels assuage their consciences with the self-deception that the young women they buy for sex are consenting adults or women who want to make a lot of money. Yet the material circumstances of the lives of the women in Cambodian closed brothels and in the Las Vegas closed Latino or Korean brothels are the same. In both places, the women are extremely poor, isolated and without family or community support, threatened by violent criminals, in captivity, raped in a pseudo-financial exchange five to twenty times a day, verbally degraded and brainwashed.

Traffickers camouflage prostitution with legitimate businesses, such as Portugese gift ceramics. Once the legitimate US business is in place, traffickers move women from Portugal, Spain or Brazil[358] to prostitute as girlfriend-for-a-week with convention-attending johns who buy a woman for the duration of a conference. The john makes his purchase via the Internet before leaving home, much like renting a car. European traffickers are able to obtain 30-day tourist visas and prostitute women in a range of US locations before returning them to their countries of origin.

Some motorcycle gangs are sex traffickers as well as pimps and drug dealers.[359] The Hell's Angels, the Outlaws, and other motorcycle gangs generate money from prostitution and trafficking in addition to methamphetamine production and sales. These gangs are involved in Nevada legal brothel prostitution as well as illegal strip club prostitution in a number of states including Nevada. The Hell's Angels, the Outlaws, and the Banditos gangs are also involved in prostitution and trafficking in other parts of the world. These gangs have fought turf wars not only in Nevada but also in other countries where prostitution is legal - the Netherlands and Australia.[360]

Las Vegas, Clark County, and federal law enforcement sources note that since 2000, Armenian organized crime has been a growing presence in escort prostitution in many U.S. cities, including Las Vegas. Armenian criminals are also attempting to move into strip club prostitution. The Armenian crime groups are exceptionally violent and are involved in identity theft and other criminal activities in addition to prostitution. Like other criminal groups, it is highly likely that the Armenian gangs use funds from prostitution to purchase weapons since they are also involved in international arms sales. Several law enforcement sources stated that there was evidence that money from prostitution is used to purchase weapons that may be later sold to terrorists.

Two non-Nevada cases provide evidence of the criminal transfer of cash from prostitution to the purchase and distribution of weapons. The first is an example of a government whose military is primarily supported with prostitution monies. The Indonesian military government has publicly acknowledged a "synergistic relationship between multinationals and soldiers in need of extra money."[361]   In her report, Trifungsi: The Role of the Indonesian Military in Business, Lesley McCulloch, a researcher for the Bonn International Center for Conversion, found that as much as 80% of the Indonesian military's budget was derived from prostitution and other illegal activities such as drug smuggling, illegal gambling, and illegal security arrangements with multinational corporations such as ExxonMobil and Freeport McMoRan.[362]

The Indonesian military became involved in illegal activities in order to provide for the costs of running the government. Major General Agus Wirahadikusumah was quoted as saying 'We all know that the military is acting as a parasite. Who backs and supports the discotheques, brothels and narcotics rings if not the military or police?"[363]

A second example of the transfer of funds from prostitution to weapons trafficking is Russian organized criminals' use of cash from prostitution for the purchase of weapons. A 2003 federal indictment charged Semyon Mogilevich with 45 counts of racketeering, securities fraud, wire fraud, mail fraud, and money laundering. Mogilevich had set up a fake Canadian company headquartered in Pennsylvania that defrauded investors in 20 different countries. He invested in illegal enterprises, including prostitution and weapons trafficking. Mogilevich is described by intelligence agencies as a gang leader of powerful organized crime groups, with operations in Russia, Hungary, Ukraine, Belorussia, Lithuania, Israel, United States, Columbia, Pakistan, Lebanon, Germany, Austria and dozens of other countries.[364]

We know that prostitution exists wherever there are military troops. And just as there is no absolute division between legal and illegal prostitution or between lap dance performance and prostitution, the legal and illegal weapons trade are not two entirely separate domains. Why should we be concerned about illegal weapons trafficking, legal arms sales, and prostitution? In today's market-driven world, the legal weapons trade and its illegal counterparts are linked to regional political and economic insecurity which in turn profoundly affect women in and out of prostitution.

The United States and Germany are the world's largest distributors of legal weapons.[365] Kathryn Farr, who has studied organized crime and the trafficking of women to the U.S., told me that she is "astounded" at the extent of the legal weapons trade in 2007.[366] Farr is currently investigating the harms to women in armed conflicts.[367] Regional armed conflicts are exacerbated by legal – as well as illegal – weapons sales and trafficking.[368] United States military training and arms sales to Africa - including Sudan in the 1980s – evolved into a disaster for the people of the Congo.[369] Floods of predators from within the Congo and even from the United Nations, descended on the vulnerable women and girls of the Congo, with countrywide devastating increases in sexual violence, including prostitution.[370] Postcolonial regional conflict, theft of indigenous resources, weapons sales, and economic vulnerability that leads to prostitution in cultures where women are devalued – are linked events.

University of Rhode Island professor and trafficking expert Donna Hughes described the use of Internet technologies by organized criminal groups to prostitute/traffic women in Las Vegas escort prostitution. The Gambino crime family gained control of telephone switching devices and rerouted phone calls from other escort agencies to their own escort operations. At the same time, some Las Vegas law enforcement officials were allegedly receiving bribes to ignore the prostitution.[371]

Organized criminals, whether as designated terrorist threats to the state, or in trafficking women for prostitution, use similar techniques in their activities.[372] Pimps and traffickers whose practices could be described as domestic terrorism of women use the same web-based technologies as those used by terrorists to expand their reach and market their 'products.'[373]

Information gleaned from the tracking of terrorist networks can and should be applied to what is being learned about sex trafficking networks. Prostitution, pornography, and trafficking networks use the same controlling and intimidation tactics, legal loopholes, and institutional weaknesses that are exploited by terrorists.[374] International feminist legal scholar Catharine MacKinnon suggested that domestic physical and sexual violence against women that is organized, systematic and international in scope, should be recognized as violation of international law.[375]

Suggesting that we in the United States should evaluate ourselves by using the same criteria used to evaluate other countries on the annual Trafficking in Persons Report, Professor Phil Williams at the Ridgeway Center for International and Security Studies concluded "The United States is a major host state to organized crime groups."[376]

8
The Role of Cab Drivers
in Las Vegas Prostitution

Taxi drivers everywhere in the world know about prostitution and procuring.[377] A number of law enforcement sources reported that organized Russian criminals were heavily involved in the Las Vegas taxi business. Several cabbies told me that approximately one man in every four who hails a cab from the Las Vegas airport or from a hotel on the Strip - asks for assistance in "getting some action" or "getting a girl." I interviewed a Las Vegas cabbie of 25 years who did not want me to use his real name. I'll call him Chuck.[378] He showed me his business card which had a cute cartoon of an old-fashioned car driving down a winding road with mountains nearby. The card advertised:

TO-GO

Gentlemen's Club Specialist

Clubs, Girls, & More....

Chuck slyly told me that he was a co-investor in or "maybe a co-owner of" an escort prostitution agency. As his business card implied Chuck pimped women to johns. His escort agency sold women to johns - the "More..." on his card - for $200. When the john paid for an escort prostitute, Chuck earned $150 of that amount. The escort agency earned $50 of the $200. The women themselves who were sold to bargain-hunting johns earned only what they were able to extract, beg, or steal from the john in tips. Chuck the cabdriver was a member of the oldest profession in the world: men who pimp women.

The economic exploitation of women in Las Vegas escort prostitution that Chuck described was more extreme than the exploitation of Ukrainian women trafficked to the United States, who were permitted to keep 30% of a $70 massage.[379]

Johns sail effortlessly from various states in the U.S. and from other countries into Nevada's legal brothels, into illegal escort and illegal brothel prostitution, and into prostitution operating in and out of Las Vegas and Reno strip clubs. The legal brothels are located in rural counties, more than an hour's drive from Las Vegas in the south, but they are only half an hour's drive from Reno in the north of the state. Illegal escort prostitution happens via mobile phone with women delivered to a john's hotel or car. Illegal brothels are usually private clubs, massage parlors or private homes.

Taxi drivers in Las Vegas have tremendous power. As one driver noted, the Las Vegas strip clubs can not be successful without cab and limo drivers, "We bring them 85 percent of their business. They can't get along without us."[380] Las Vegas has the fourth busiest airport in the United States and it depends on cab drivers. After people receive free drinks from casinos, they can not drive.[381]

Chuck delivered johns only to those strip clubs that paid an extortion fee upon delivery of each customer. Several cab drivers told me that they obtained fees averaging $70-90 per customer. In 2004, Spearmint Rhino (a strip club) in Las Vegas was estimated to have paid $4.8 million in extortion to cab drivers.[382] That same year, KLAS-TV reporter George Knapp estimated that Las Vegas strip clubs paid a grand total of more than $20 million to taxi drivers, with individual drivers sometimes averaging $1000 in kickbacks from one club per weekend night. Strip clubs were threatened with a total cab boycott unless they continued to pay.[383] In 2007, the estimate of total cabdriver extortion fees, none reported to IRS, was upwards of $50 milion.[384]

In 2006, Chuck was angry at Spearmint Rhino because they were refusing to pay a kickback, so he and other drivers diverted customers from that club (and others that refused to pay up), recommending other strip clubs to customers. Chuck's approach was to recommend only the Las Vegas strip clubs that paid him at least $50 per customer.

Las Vegas cab drivers accuse hotel doormen of pimping, and vice versa. When doormen learn that hotel customers are heading for strip clubs, they divert customers from a cab to a limousine company, in return for a kickback from the limo driver, cutting taxi drivers out. Half of the strip club extortion fee thus ends up with the doorman rather than cabbie.[385]

The Las Vegas cab drivers resent limousine drivers who take customers from downtown Las Vegas (Clark County where prostitution is illegal) to the nearest legal brothel in Nye County, about an hour and a half's drive. When Chuck drove Las Vegas johns to the legal brothels, a percentage of the brothel's prostitution fee was set aside for each customer he delivered. For example, Sheri's brothel paid him 30% and Chicken Ranch brothel paid 40%. The remaining amount of the john's payment was split between the brothel owner/pimp, the brothel bartender, brothel madam/assistant pimp, floor managers, and other junior pimps. The women were doing well if they earned anything more than 30% of the john's payment.

The run from Las Vegas to the Nye County legal brothels sometimes brings enormous profits. For example in 2005 Chuck shuttled a group of twelve U.K. prostitution tourists from Las Vegas to a Nye County brothel. The men were charged $2700 per person for an afternoon "date" with a prostitute, or $39,000 for all 12 johns including tips. Chuck rented a van and earned $15,000 from that one trip.

Johns are delivered by taxi drivers not only to legal brothels in nearby counties, and to Las Vegas stripclubs, but also to illegal escort and massage prostitution. The kickback scheme remains in place whether or not the prostitution is legal or illegal.[386]

Some of the Las Vegas stripclubs - Cheetahs, Crazy Horse Too, Sapphire, Treasures, Spearmint Rhino and Scores - met in 2005 to discuss resisting the kickbacks to cabbies. The Las Vegas City Council has struggled to address the conflict between taxi drivers and strip clubs. When the City Council attempted to block the extortion fees demanded of stripclubs by cab drivers, the drivers threatened to block the road from the airport to the Strip – striking fear into the hearts of casino and hotel owners and politicians alike. In 2005, a bill was introduced in the Nevada legislature that would have prohibited cabdriver extortion. Governor Kenny Guinn vetoed the bill.[387] This legislative support for extortion, as writer Steve Miller described it, was reaffirmed in 2007 when the entire Nevada legislature voted again to uphold Guinn's veto of a law that would ban extortion by cab drivers.[388]

"Cab drivers are pimps," a strip club bouncer told me.[389] This example of pimping in Las Vegas by taxi drivers is not meant to exclude others who are also pimps: casino hosts, receptionists, hotel managers, airport employees, travel agents, restauranteurs, strip club employees, valets, limo drivers, tour guides and others. The pimping income - or tipping as it is often described in Las Vegas - is money obtained from the sexual exploitation and sexual abuse of women in prostitution. If the definition of pimping is someone who finds customers for prostitutes and profits from that, then there are many pimps in Las Vegas.

9
Political and Judicial Corruption
and the Nevada Sex Industry

Corrupt police, politicians and bureaucrats give the industry their unofficial blessing and protection. In return they receive bribes, political support and sexual favors…. In a number of places the police do not tackle the traffickers and the brothel owners because they *are* the traffickers and the brothel owners.[390]

The author of this observation was writing about trafficking of women into prostitution in India, but much of it applies to Nevada. Sex businesses are integrated into the legal and political power structure in Nevada just as they are in India. Trafficking of people into a sex industry as large as that in Las Vegas can happen only with the involvement of government officials. The corruption of federal as well as state and municipal public officials is necessary to the success of organized criminal operations in for example, trafficking women from Korea into Las Vegas massage parlors or trafficking women from Oregon to Las Vegas strip clubs. Several members of the Indian Parliament were arrested in 2007 for their facilitation of the trafficking of Indian women via criminal collaboration with travel agents and with those who produced fake visas and other legal documents.[391]

Evidence from many sources suggests that large amounts of cash from pimps, brothel owners, strip club owners, and taxi drivers, among others, keeps some Nevada politicians in office. In Nevada, some politicians can be bought for $5-10,000.[392] A Nevada pimp proudly explained to the author, "We know a lot of people in the governor's office. We contribute to both Democratic and Republican campaigns in [closest city to this particular brothel]." He also pointed to a framed letter in the brothel waiting area, a letter from a former US president, thanking him for a campaign contribution.

In 2006, a Nevada Secretary of State running for another state office accepted campaign contributions from brothel owner and northern Nevada developer Lance Gilman and also from the owner of Sin, a strip club in Las Vegas.[393]

The complicity of Nevada county sheriffs, district attorneys, and judges in legal prostitution has been exhaustively and brilliantly documented by Jeanie Kasindorf.[394] Today, Nevada politicians are "just as buyable as they were 30 years ago."[395] Nonetheless, some politicians resist. In 2006, Nye County Commissioner Candice Trummell assisted the FBI in its prosecution and indictment of a pimp who attempted to bribe her.[396] Steve Miller was a Las Vegas City Councilman from 1987-1991, during which time he refused to accept a bag containing $10,000 cash offered to him by an escort agency owner.[397]

Political influence and other corruption in Las Vegas occur via campaign contributions, consulting fees, acts of prostitution, and bribes in exchange for favors, zoning changes, licenses, and permits. In the 1980's, an FBI undercover operation resulted in indictments of five politicians, and in the extortion conviction of Nevada State Senator Floyd Lamb.[398]

"To understand how the process of political corruption operates," stated a Las Vegas police officer, "Go to the Las Vegas city council meetings, and watch the zoning permit process." The following description by Brent Jordan confirms the accuracy of the officer's observation. Jordan described an FBI operation that exposed bribes of city officials by organized crime in strip clubs:

> ...multiple current and former Las Vegas City Council members... were receiving cash pay envelopes weekly... Watching an ethics-challenged [Las Vegas] councilman approach the back door of [Cheetah's strip] club's office dressed in an over sized coat and baseball cap pulled down over his eyes, as if the disguise might help conceal the features that make the local news nightly, to receive a cash "campaign contribution" would be laughable if not for the fact that Cheetah's concurrently received an okay for another lucrative and coveted topless club location during the following City Council meeting on these matters.[399]

Jordan's account was validated by court testimony of Michael Galardi, owner of Cheetah's strip club in Las Vegas, who stated that

> ....he initially paid off elected officials so they would protect his strip club empire from harmful legislation. He feared that if he stopped payments, politicians might retaliate by introducing stricter regulations on the clubs, he said.

> Former Las Vegas City Councilman Michael Mack was a perfect example, Galardi said. Mack acted like he owned Cheetah's, Galardi's strip club in Las Vegas, and would command club managers to "bring me your best whores," Galardi testified. City Councilman Mack told Galardi he would push for an ordinance that required dancers to perform six feet away from patrons if Galardi stopped paying him or stopped providing him with women who would perform sex acts on him. "We got tired of being strong-armed by Mr. Mack, so we decided it was time to play dirty back," Galardi testified.

> Managers at Cheetah's captured security footage of Mack receiving oral sex from a stripper. Galardi saved the video and said he planned to release it to the media if Mack introduced his ordinance. A video labeled "videotape of Mike Mack" was listed among items seized by FBI agents in their May 2003 raid of the strip club.[400]

Galardi testified that a second Clark County Commissioner, Dario Herrera, also received monthly cash payments and was provided free lap dances and oral sex at Cheetah's.[401] Gallardi testified that he also bribed a County Business Licensing Director and a County Manager.[402]

Four of seven Clark County commissioners were federally indicted in 2006 for corrupt practices that were linked to the Nevada sex industry.[403] Clark County Commissioner Erin Kenny admitted to corruption, including a lengthy series of votes on behalf of the interests of Cheetah's strip club owner Mike Galardi such as favorable zoning for Cheetah's and unfavorable votes for Cheetah's competitors. She consistently voted against ordinances that would ban sexual touching in clubs.[404] For each vote, Galardi paid

County Commissioner Kenny $5000-10,000, totaling at least $50,000. Kenny also accepted $100,000 from a Las Vegas real estate developer.[405] She was charged only with wire fraud in exchange for her testimony against two other county commissioners who had accepted similar bribes.

International trafficking researcher Donna Hughes, using data from Brunon Holyst, described the influence of organized crime on political institutions in former Soviet states. Although Hughes was writing about Ukraine, her analysis also applies to the corruption of Nevada public institutions.

> As the influence of criminal networks deepens, the corruption goes beyond an act of occasionally ignoring illegal activity to providing protection by blocking legislation that would hinder the activities of the groups. As law enforcement personnel and government officials become more corrupt and members of the crime groups gain more influence, the line between the state and the criminal networks starts to blur.[406]

A 2006 Los Angeles Times investigative report by Michael Goodman and William Rempel detailed extensive judicial corruption in Nevada, naming five district judges who accepted campaign contributions that compromised their judicial impartiality but which were accepted because, several said, it was 'business as usual.'

> During the most recent Nevada election in which all district judgeships in Las Vegas were on the ballot, 17 incumbents raised more than $1.7 million in campaign funds, collecting much of it from lawyers and casinos with cases pending before them, campaign financial reports and court records show. At least 90% of all contributions for the election, held November 5, 2002, came from lawyers and casinos.

> Frequently, a donation was dated within days of when a judge took action in the contributor's case, the records show. Occasionally the contribution was dated the same day. [407]

In some notable cases, judicial and political corruption has facilitated organized crime as it operates in Las Vegas strip clubs. For example, Treasures stripclub in Las Vegas is owned by Houston, Texas brothers Ali and Hassan Davari who also co-own Texas strip clubs. Treasures clubs in both states have had multiple prostitution convictions, with money laundering investigations in Texas.[408] A district court judge who accepted campaign contributions from people with a vested interest in the outcome of the decision later dismissed a licensing challenge to the club when a prostitution solicitation conviction occurred. This appears to be judicial corruption.[409]

Many of the Las Vegas strip clubs have contributed large sums to judges' election campaigns. This pattern of organized crime is not unlike the way that organized crime is closely connected with the judicial system and with politicians in other countries. For example in India, a member of Parliament and his aide were charged in 2007 with assisting traffickers in evading immigration laws when they trafficked women and children out of India.[410] Votes from pimps, organized criminals and women in prostitution have been offered in exchange for protection and support of organized crime.[411] And in Mexico, a federal police officer told Peter Landesman that high level government officials in the state of Sonora received $200,000 weekly payoffs from traffickers. "Some officials are not only on the organization's payroll, they are key players in the organization," an official at the U.S. Embassy in Mexico City told Landesman. "Corruption is the most important reason these networks are so successful."[412]

In Mexico, when a woman was beaten, raped, and her home threatened with being burned down, no judge was willing to hear the case against her perpetrator.[413] Yet when a man stole a pack of cigarettes in a Walmart, the case was immediately brought before a judge. Because of these repeated judicial refusals to understand the context of violence in which women exist and because their voices are sometimes not even allowed into the courts, Mexican women deeply mistrust the legal system. Women prostituting in Las Vegas have similar fears that their legitimate concerns about exploitation and violence will continue to be contemptuously dismissed.

For nine years, former Las Vegas City Councilman and national crime columnist Steve Miller tracked the illegal practices, political payoffs, and violence at the Crazy Horse Too strip club in Las Vegas. Johns at the club were scammed out of thousands of dollars each when they were solicited for prostitution during lap dances (which are frequently prostitution), having paid in advance for prostitution with credit cards. Miller documented organized criminal involvement in the Crazy Horse Too club: extortion, beatings, murders, prostitution - all occurring with the apparent knowledge of city officials, judges, and law enforcement. Miller and others have documented payoffs to police officers, district attorneys, and judges.[414] Las Vegas strip club owner Rick Rizzolo and his close associates contributed more than $135,000 to many public officials. See Table 5.

The practice of political and judicial payoffs continues, not only in Las Vegas, but also in Seattle, where the owners of 4 stripclubs were charged with 29 counts of illegal campaign contributions to City Council members. When crucial zoning decisions were up for a vote, just as in Las Vegas, the Seattle stripclub owners made thousands of dollars of 'campaign contributions' which functioned as bribes.[415]

When the very occasional politician, lawyer, or other public figure refuses a bribe from sex racket businessmen, there can be hell to pay. Verbal harassment, intimidation, and unsolicited checks in the mail from unknown parties have been reported by politicians and other public office-holders who refuse bribes.

In addition to corruption on the part of public officials and judges, there are also documented and many undocumented instances of police exploitation of women and children in Nevada prostitution, usually extorting acts of prostitution in exchange for non-arrest.[416] In a case of law enforcement corruption, a Las Vegas police specializing in cybercrimes was disciplined and transferred after an internal probe found that he accepted $15,000 from Rick Rizzolo, a strip club owner.[417] Unfortunately, it takes only a few corrupt police officers to tarnish the reputation and the effectiveness of the entire force.[418]

Table 5. Cash Contributions to Public Officials by a Las Vegas Strip Club Owner*

Mayor Oscar Goodman	$40,000
District Attorney David Roger**	13,000
City Councilman Gary Reese	10,000
District Judge Joseph Bonaventure	6,000
City Councilman Michael Mack	6,000
District Judge Donald Mosley	6,000
Las Vegas Municipal Judge George Assad	5,000
Las Vegas Municipal Judge Bert Brown	5,000
Las Vegas Municipal Judge Toy Gregory	5,000
District Judge Sally Loehrer	5,000
District Judge David Wall	5,000
District Judge Jessie Walsh	5,000
District Judge Lee Gates	5,000
Clark County Clerk Shirley Parraguire	2,500
District Judge Nancy Saitta	1,000
District Judge Michael Cherry	1,000

* This table is from a graphic constructed by Mike Johnson and Michael Squires for the Las Vegas Review-Journal, January 20, 2005.[419]
**Roger returned this contribution after he was criticized by a political opponent.

A reliable source informed us that in the early 1990s, the Las Vegas Metro Police Vice Department's policy was to conduct highly abusive "room setups" in the Strip hotels. In these prostitution arrests, undercover police officers solicited prostitution from an escort service, with additional officers hidden in an adjacent room. When the women's clothes were removed in preparation for prostitution, the arrest was made. At that time, the policy of the police department was to interview the women while they were naked, including photographing them while naked. The stated purpose of this policy was to deliberately humiliate and psychologically break down the arrested woman so that she would give up the name of her pimp.

This policy is apparently not in effect now but while it was in effect it clearly violated the women's civil rights and may even have met international legal standards that define torture. Torture is

> ...any act by which severe pain or suffering, whether physical or mental, is intentionally inflicted on a person for such purposes as punishing him... or intimidating or coercing him or a third person, or for any reason based on discrimination of any kind, when such pain or suffering is inflicted by or at the instigation of or with the consent or acquiescence of a public official or other person acting in an official capacity."[420]

The Istanbul Protocol is an international document that defines what torture is. It lists specific acts that are considered to be torture. These acts include verbal sexual harassment, *forced nudity*, rape, *sexual mocking*, physical sexual harassment such as groping, and not permitting basic hygiene[421] It is clear that the Las Vegas Metro police policy at one time included forced nudity and sexual mocking, acts that fall within the Istanbul Protocol's definition of torture.

In 2007, psychiatrist Metin Basoglu and his colleagues released a study demonstrating that *mental* torture resulted in the same suffering and the same long term emotional harms as did *physical* torture.[422] This Las Vegas police department policy was tragically typical of other police practices in the United States that treat women in prostitution as less than human and that violate their rights to be treated with dignity.

These are only a few of many instances of money from criminal enterprises being funnelled into public institutions, corrupting them. Cash is sometimes diverted from legitimate public institutions into criminal enterprises. For example tax preparation companies have reportedly provided loans to pimps in order to pay for escort advertising in the Las Vegas Yellow Pages.[423]

Intrepid writers like Steve Miller, John L. Smith, and more recently Joshua Longobardy in Las Vegas keep the public informed about these issues - intimidation, death threats and defamation lawsuits notwithstanding.[424] Attempted and sometimes successful censorship of reporting on prostitution, sex businesses, organized crime and trafficking happens in many parts of the United States. For example, a reporter in South Carolina was fired when he insisted on reporting a story about an illegal brothel that was favored by a local sheriff, a newspaper managing editor, and the editor's political cronies.[425]

10
It's the Advertising, Stupid!

How did Las Vegas become a destination for prostitution tourists and a city to which women from all over the US and the world are trafficked? Although prostitution is not legal in Las Vegas, you wouldn't know it. It is openly and globally advertised.[426]   Slyly lifted from the Alcoholics Anonymous caveat about confidentiality in AA meetings ("what happens here, stays here") the Las Vegas Convention and Visitors Authority (LVCVA) advertising slogan "what happens in Las Vegas stays in Las Vegas" is a coy hair's breadth away from openly promoting prostitution.[427] At least one pimp was very clear about "what happens in Las Vegas stays in Las Vegas" means despite hypocritical disclaimers issued by politicians and casino hosts that the ad campaign has nothing to do with prostitution. "Don't you know what that means around the country?" he asked one reporter.[428] Well, yes, lots of people know what it means: women and girls are for sale in Las Vegas and your wife or girlfriend won't find out about it when you go home.

An LVCVA-sponsored radio dialogue featured a woman suspiciously asking her boyfriend what he did in Las Vegas. Answering in an evasive, guilty manner he replies, "Uh... we ate?" She presses him further, and he responds "Honey, it was a men's weekend, we talked and stuff." The ad ends with the LVCVA's editorial comment: "Our fine dining could be your alibi."[429] The LVCVA's website offers an "alibi kit webpage where you can change your identity for your Las Vegas trip.[430]

The Hard Rock Club in Las Vegas featured a billboard with a naked woman lying on a blackjack table with a card in her mouth, above the words "There's Always A Temptation to Cheat."[431] Billboards and club names in Las Vegas promote prostitution as game hunters' adventures with strip clubs named Cheetah's, Jaguar's, and Jungle Room. Another strip club advertised on a billboard along Interstate #15, "Where Fantasy Becomes Reality."

Mainstreaming prostitution via technology is the annual AVN Adult Video Entertainment Expo which meets in Las Vegas. A comparable group in Australia, where prostitution is legal, similarly brings together sex business entrepreneurs, lobbyists, pimps, and some politicians. These public relations groups serve to mainstream prostitution and pornography via sex industry exhibitions.[432] Las Vegas entrepreneurs promote the appearance of sexy sinfulness but avoid mention of the city's annual "Pimps Up - Hoes Down" convention which would too bluntly expose the abuses and brutal power imbalance of prostitution.

We wanted an empirical measure of how successful the city fathers and the Convention and Visitors' Authority of Las Vegas had been in the wink-wink-it's-here-but-you-don't-really-see-it prostitution advertising campaign. In December 2005, five interviewers surveyed 154 people walking on the Strip where many well-known hotel/casinos are located. Lots of prostitution transactions happen on the Strip and in the hotels. We asked these folks, "Is prostitution legal in Las Vegas?" Sixty percent of our respondents were male and 40% were female. Most respondents were from states other than Nevada; somewhat fewer respondents were from outside the United States.

Responses to "*Is prostitution legal in Las Vegas?*" Asked of 154 men and women on the Las Vegas Strip		
Yes	44%	(67)
No	53%	(82)
Unsure	3%	(5)

Since 44% of the people we surveyed believed that prostitution is legal in Las Vegas, we concluded that the Las Vegas Convention and Visitors Authority advertising campaign had successfully telegraphed the message that prostitution was available in Las Vegas despite its illegal status in Clark County. Respondents were asked "Where did you receive the information that prostitution is legal or illegal in Las Vegas?" They told us that they heard about it on the Internet, from the business cards on the Strip that advertised prostitution, on TV or other news media, and via word of mouth. One respondent observed that he concluded prostitution is legal in Las Vegas because it is not prosecuted.

The disinformation regarding prostitution's legality in Las Vegas has traveled as far as Macau, where a 2007 report accepted the myth that prostitution is legal in "Sin City."[433] This myth is accepted in Macao, which is both a major hub of sex trafficking, and the new Asian location of Nevada casino entrepreneurs' expansion.

Advertising for prostitution does not distinguish between domestically or internationally trafficked women. The industry is focused on market demand. When responding to prostitution advertising, predatory johns usually have no awareness that they may be buying trafficked women or minors for sex - but in fact they frequently are doing just that. It is impossible for the john to know whether a woman has been trafficked from another city, another state, or another country. Under duress from pimps or traffickers, women hide their coerced status in prostitution. A recent bust of a brothel in Las Vegas highlights this dilemma. Even when police arrived at an illegal brothel, women who probably had been trafficked told police that they were there voluntarily, probably following traffickers' instructions to tell the police: "I'm happy, I'm making money."[434] Having experienced multiple betrayals, trafficked and prostituted women trust no one, including police and advocates.

Trafficked women are sometimes advertised for prostitution. Three law enforcement agencies and the Internal Revenue Service busted an illegal brothel in Las Vegas where women were being prostituted by suspected Chinese traffickers. A Las Vegas Metro Police officer stated that the traffickers "advertised with business cards handed out to tourists and through taxi drivers, who brought customers in exchange for kickbacks."[435] An escort prostitution review on the Internet described an agency named "Zuzana" located near the airport that *"rotates ladies of eastern European origin into Amsterdam for brief periods."*[436] This advertising hints that the women at

Zuzana are very likely to be eastern European women who were trafficked into Dutch prostitution.

Prostitution is promoted in most US cities by classified advertising in daily or weekly or online newspapers and also telephone directory advertising. Prostitution is named entertainment or massage. In fact, the ads should read: "Women for rent or sale. Choose by race, age, disability, or country of origin. Do whatever you want to them and call it sex whether they are willing or not. Don't delay – enjoy a brothel today!"[437]

We calculated how much money was spent on advertising in order to understand the success of the sex industry in Las Vegas. Las Vegas has more prostitution telephone directory advertising than any other US city: 165 pages in 2006, 173 pages in 2007. The cost of the Yellow Pages advertising for "Entertainment" and "Massage" which are in reality prostitution was conservatively estimated at $12.4 million per year. We also estimated costs of prostitution advertising on the Strip. We estimated that it cost more than $11 million to produce the flyers and pamphlets in newsracks and the business cards thrown everywhere on the Strip each hour of the day. We further estimated nearly $3 million in costs of prostitution advertising on the Internet. **A conservative estimate of costs for these four types of advertising for prostitution is more than $26 million per year.** See Table 6.[438] The advertising message is that prostitution is readily available and legally tolerated in Las Vegas.

We looked closely at the advertising for prostitution in Las Vegas and its obvious connection with who is trafficked into prostitution. In 1999, Thailand, Vietnam, China, Mexico, Russia, Ukraine, and the Czech Republic were primary source countries for trafficking of women into the United States.[439] Women of these ethnic backgrounds are prostituting in both legal and illegal prostitution in Nevada. In fact, many of them are advertised by ethnicity for sale in the Las Vegas Yellow Pages.

Annie Fukushima collaborated in our analysis of Yellow Pages advertising for prostitution in Las Vegas. There are a total of 1387 advertisements for prostitution in the January-July 2007 Las Vegas Yellow Pages. These advertisements are on 173 pages of the telephone book. All ads for prostitution are categorized as "entertainment" and "massage."[440] A number of the advertisements suggested that those prostituting were young or students. Women and youth were advertised as Teens, Wild Teens, AAA-Full Service Teens, Wild Barely Legal Briana & Haily, Just Out of High School, Barely Legal, Petite Japanese Teen, College Girl, Cheerleaders, School Girls, Barely Legal School Girls Cutting Class, Barely Legal China Doll, Exotic Student Nurses, Students, Exotic Teens, University students. "Mature women" were for sale at cheaper rates, advertised as "affordable."

The word "exotic" is code for women of color. Racism was apparent throughout prostitution advertising in Las Vegas. The derogatory term "Oriental" was regularly used in connection with Asian escort or massage prostitution, for example: Asian Discount Teens, Chinese Take Out Direct to You, Full Service Orient Express, Lovely Orientals, and Asian Buffet. Annie Fukushima pointed out that prostitution advertising in Las Vegas condescendingly portrays Asian women as perpetual outsiders. Some Las Vegas Yellow Pages advertisements eroticize the vulnerability of English as a second language in the following: an 18 year-old Korean girl is advertised as saying in broken English, "No Happy – No Pay" or a Vietnamese young woman informs potential johns, "I stay long time." Asian women advertised for sale by ethnicity in prostitution include Thai, Chinese, Japanese, Korean, Cambodian, Eastern Asian, Filipino, Hong Kong, Indian, Malaysian, Singaporean, Taiwanese, Tokyo, and Vietnamese. Dismissing the women's actual ethnic identities (because that identity is unimportant to the johns who buy them), a confusing hodgepodge of Asian women were offered for sale: for example China Geisha Girls or Thai/Filipino.[441]

Other racial or ethnic identities specifically advertised to johns were Black Women, Black Fantasies, Black Diamond Dolls, Beautiful Black College Girls, Brazilian Women, Exotic Brazilian Beauties, Middle Eastern Women, Latinas (usually described as Latinos), Indigenous, Latin Fantasies, Spicy Latin Girls, Spicy Spanish Girl, Petite Latino Playmate, Wild Latino Selena.

Apparently some johns' fantasy is the brown undocumented immigrant. See for example, the 8th street Latinas website which advertises, "See hot, young & brown Latinas that will do absolutely anything to get their citizenship!"[442] Also advertised for sale were Full Service Blonds, Busty Blondes, Swedish, Russian, Swiss, Persian Beauty, Fiery Irish Redhead.

Multimillion dollar advertising campaigns in all media send the message that prostitution is acceptable and tolerated in Las Vegas. The citizens of the state look the other way.

Table 6. 2007 Estimated Advertising Costs for Prostitution in Las Vegas *

	Cost per month	Cost per year
Print Advertising	**$921,183**	**$11,054,195**
Newsprint Catalogs 2663 magazine racks containing estimated 2,263,550 catalogs at a cost of $0.331 each plus newsrack annual fee of $25 each /year	754,783	9,057,395
Color Full Page Flyers 1 million 2-sided color flyers per month at $0.137 each	137,000	1,644,000
Escort Cards 2 million 2.5 x 4 inches/month at .0147 each	29,400	352,800
Yellow Pages Phonebook 173pgs	**$1,053,433**	**$12,641,196**
Full page and 3/4 page size 128 ads @ $3,766-6,641 /month	713,011	8,556,132
Half page ads and smaller size 410 ads @ $605-4,713 /month	340,422	4,085,064
Internet Advertising	**$235,325**	**$2,823,900**
3 directory ads plus the cost of a small website		
TOTAL ADVERTISING FOR PROSTITUTION IN LAS VEGAS	$2,209,941	$26,519,292

*See Appendix B for additional details.

11
Pornography, Prostitution, and Trafficking in Nevada

"What happened to common sense?" asked Roger Young, retired Nevada FBI agent. "The fact that there is a camera filming the prostitution doesn't change the fact of the prostitution. Pornography is essentially a crime scene surveillance tape. You can't say to someone, hey let's go rob a bank but if we film it then it won't be robbery."[443]

Pornography was described by an editorial in The Economist as the "product side of the sex business."[444] The women whose prostitution appears in pornography are prostituted women. Pornography is also advertising for prostitution. Men learn how to use women by looking at and masturbating to pornography, often developing a taste for prostitution.[445] In order to conceal the harms that are documented in the picture, the pornographer disconnects the picture from the person. The pornographer and his allies then name what is happening to her in the picture, "adult entertainment," or "free speech" rather than "sexual predation" or "torture."

Since 1985, production of pornography has greatly expanded. Themes of degradation, humiliation and violence against women have in effect colonized U.S. popular culture.[446]

Pornographers are indistinguishable from other pimps.[447] Both exploit women's and girls' economic and psychological vulnerabilities or coerce them to get into and stay in the sex industry.[448] Pornographers and pimps both take pictures to advertise their "products," suggest specific abuses for johns to perpetrate against women, and minimize the resulting harms. One pornographer advertised that he was in the business of "degrading whores for your viewing pleasure," clearly eliminating any boundary that is alleged to exist between pornography and prostitution.[449]

Women have explained that they study pornography in order to learn how to perform prostitution: "I watch pornos and act like that in the room [with a john]."[450] Men show pornography to women to illustrate what they want them to do. Strip clubs show video pornography to promote lap dance and VIP-room prostitution. Legal brothels show pornography in the waiting area to speed men along in their purchases of women.

The filming of 251 men's prostitution of Grace Quek (called Annabel Chong) was sold as pornography named "The World's Biggest Gang Bang."[451] After being edited down to 4 hours, the film became hardcore pornography.[452] The filming of johns assaulting Quek was stopped after 10 hours because she was bleeding internally. For Quek, the film was not just acting, the film was not just a representation of rapes. Instead, it was real: real johns perpetrated real sexual assaults on her resulting in real physical and psychological injuries.

Organized criminals have always had a major stake in pornography production, just as they are heavily involved in non-filmed prostitution. Michael Zaffarano, a member of an organized crime group, was a strip club owner and a producer and distributor of pornography in the 1980s.[453] Jenna Jameson, a pornography star, grew up in Las Vegas and entered the sex industry at a strip club when she was age 16, removing the braces from her teeth in order to get hired by a strip club that was owned and managed by organized criminals. Jameson quickly moved into prostitution and pornography production.[454]

Survivors of prostitution and johns alike explain that pornography is prostitution with a camera. One john explained, "Yes, the woman in pornography is a prostitute. They're prostituting before the cameras."[455] A number of courts have understood that making pornography is an act of prostitution.[456] Pimps make more money from johns when they advertise women in prostitution as "adult film stars" who are available as "escorts."[457]

In Nevada, it's impossible to separate pornography production from prostitution and trafficking. Visual pornography is a record of prostitution or trafficking. Pornography is a documentary of specific women's abuses in prostitution, and its consumers obtain pornography as a filmed document of someone's sexual humiliation.[458] Pornography is a document of what men's domination of women in prostitution looks like - in all its racist and classist specificity. Today, pornography is one way to traffic women.[459] On pornography/prostitution websites, women are for rent and sale. They are moved across town, across the country, and from one country to another.

There is a crossover of people involved in legal prostitution to illegal prostitution and back again. Similarly, web-based, video, and print pornography are inseparable from the rest of the sex industry, with the same kinds of crossovers from prostitution to pornography to sex trafficking. Some Nevada legal brothel pimps have declared their economic interest in "cross fertilizing" prostitution with other legal adult businesses such as strip clubs, Internet sex sites and pornography.[460] "The Girls of Cheetah's" is pornography made at a Las Vegas strip club.[461] Articulating the connection between stripping and pornography, one strip club website advertised, "Breeding pornstars: one showgirl at a time!!!"[462]

The Internet has expanded the reach of traffickers and it has intensified the humiliation and violence of prostitution. Craigslist for example is an Internet site where people can post at no cost what they want to buy and what they want to sell. A cell phone and a free ad on Craigslist are all that a pimp needs to run a teen or adult in prostitution. In March 2005, Craigslist in the United States averaged 25,000 new ads every 10 days for "erotic services" that are probably prostitution. A war between pimps and flaggers (those who go online to warn about or flag those who appear to be pimps advertising on "women seeking men" or "men seeking women" online prostitution sites) was raging on the Las Vegas Craigslist site in 2007.[463]

Distinctions between various arms of the sex industry are blurred, especially when a previously marginal segment of the industry such as online prostitution, is mainstreamed.[464] The web technology of live video permits johns to obtain prostitution online which can not be distinguished from trafficking and filmed sexual assaults of children. Writing in 2004, Peter Landesman described the evolving sameness of internet pornography, prostitution, trafficking and slavery. In this example, these crimes can not be differentiated.

> Immigration and Customs Enforcement agents at the Cyber Crimes Center in Fairfax, Va., are finding that when

it comes to sex, what was once considered abnormal is now the norm. They are tracking a clear spike in the demand for harder-core pornography on the Internet. "We've become desensitized by the soft stuff; now we need a harder and harder hit," says I.C.E. Special Agent Perry Woo. Cybernetworks like KaZaA and Morpheus / through which you can download and trade images and videos -- have become the Mexican border of virtual sexual exploitation. I had heard of one Web site that supposedly offered sex slaves for purchase to individuals. The I.C.E. agents hadn't heard of it. Special Agent Don Daufenbach, I.C.E.'s manager for undercover operations, brought it up on a screen. A hush came over the room as the agents leaned forward, clearly disturbed. "That sure looks like the real thing," Daufenbach said. There were streams of Web pages of thumbnail images of young women of every ethnicity in obvious distress, bound, gagged, contorted. The agents in the room pointed out probable injuries from torture. Cyberauctions for some of the women were in progress; one had exceeded $300,000. "With new Internet technology," Woo said, "pornography is becoming more pervasive. With Web cams we're seeing more live molestation of children." One of I.C.E.'s recent successes, Operation Hamlet, broke up a ring of adults who traded images and videos of themselves forcing sex on their own young children.[465]

Las Vegas hosts a large number of Internet web cam pornography sites.[466] On a web cam site the john pays to chat with live women who perform prostitution on live streaming video, performing what the johns pay them to do. The webcam company advertises via a network of affiliates who post ads on their websites. The affiliates earn a fee whenever someone signs up to purchase webcam pornography.

A law enforcement investigation in Las Vegas located a multi-use sex industry operation that included online prostitution. Looking like a small office complex from the street, it blended pornography production with escort prostitution. The pimp/pornographer rented 5-6 offices on Tropicana Avenue which functioned simultaneously as Internet pornography, cyber-peepshow prostitution, and a location out of which women were pimped to Las Vegas hotels and to an illegal brothel.[467]

Web cam video and escort prostitution sites have recently merged with adult dating sites. Since 2000 there has been increasing prostitution advertising on dating websites with the major dating websites now largely consisting of locations where johns seek women who they presume to be prostituting. Adultfriendfinder for example features gonzo pornography[468] of women seeking dates for prostitution in dozens of countries and every state in the United States. The site is available in German, Spanish, Japanese, French, Portuguese, Italian, Dutch, and Swedish.

It has been established by European intelligence agencies that trafficking occurs via websites where prostitution is advertised.[469] Many dating websites are headquartered in Nevada, thereby playing a role in online trafficking of women.

According to investigative reporter John Smith, the southern California pornography industry has established a foothold in southern Nevada. Major production of pornography is occurring in Nevada, with some pornographers purchasing legal brothels.[470] Pornographers and other pimps hold conventions in Las Vegas where pornography production has moved some of its operations from its base in the San Fernando Valley, California.[471]

> It's only natural that the denizens of the porn world would eventually begin migrating to Sin City… the XXX movie industry and Vegas go hand-in-hand. Twice a year, Adult Video News, the pornography trade magazine, holds a conference in Las Vegas.[472]

Like Smith, reporter Abowitz describes "porn roots" being established in Las Vegas. Ray Pistol's Las Vegas based company, Arrow Film and Video, is at the center of the expansion of Nevada's pornography production. Pistol at one time rented a casino in Las Vegas to film pornography.[473] Thomas Zupko,[474] a producer of films of men anally raping women, has a business association with Ray Pistol in Las Vegas.[475]

The sex industry in Las Vegas is expanding in size and in the diversity of the venues where it is operating. Pornography plays a crucial role in the expansion of these businesses of sexual exploitation.

12

Barriers to Services for Women Escaping Nevada Prostitution and Trafficking

Jody Williams

If you've ever watched a daytime talk show you've probably seen a woman talking about what her life as a prostitute was like- drugs, alcohol, violence, loss, trauma, death, and suffering. You've probably wondered why doesn't she just leave? But why she "chooses" such a life isn't the real question. The real question is "what's stopping her from getting out of there?" If you interview prostitutes, you'll find almost all of them have tried unsuccessfully to quit many times. We have all the tools we need to assist these women to get out of prostitution - but there are some barriers they can't cross to access this assistance. These barriers include the following.

Barriers to Obtaining Shelter for Women Escaping Prostitution in Nevada

In order to leave prostitution, women must make a change of housing or give up where she is currently living. While some programs offer shelter, the woman escaping prostitution might not meet eligibility requirements for entry. For example, most programs will not take children, pets, women who are HIV positive or women who have communicable diseases such as tuberculosis, or transgender prostitutes. Many women in prostitution with children are not going to leave without taking her children with her.

Some shelters will not admit a woman who has not been drug free for a predetermined amount of time. Other programs will not admit women who have recently been released from prison. Some housing programs won't admit women with criminal records.

Other shelters will not accept a woman if she does not conform to the shelter's own religious tradition. In one unfortunate case, a woman who was a Buddhist of 20 years was told that she had to discontinue chanting and meditating and replace her Buddhism with Christianity. She was told that she'd never escape prostitution and get sober unless she converted. This is unacceptable on the grounds of religious freedom but it also reflects shelter staff's ignorance about pimps' mind control and the sensitivity of prostituted women to being told how to think. A Christian shelter in Thailand, on the other hand, had pictures of many religious icons, including Buddha, Jesus, and indigenous women healers, on its chapel walls - sending the message that although the shelter was run by nuns, they welcomed all religious traditions.

Many women are lost because of the lengthy waiting period required in order to receive services. After being put on a waiting list, she might not be reachable at the moment when space is available, and she then loses the opportunity. If the waiting period is too long, by the time she is told she can enter the program, she has gone back to the streets to survive. When she goes back to prostitution, she is surrounded by people who encourage her to continue prostituting, including pimps who prevent her escape.

Another reason why women may avoid seeking shelter is because they fear physical and/or sexual violence at the shelter. It is not uncommon for women to be beaten and/or raped by other residents or even staff. There is a need for women-only shelters staffed by women.

Women who have outstanding warrants are afraid to seek shelter for fear of arrest. In addition, if she is attempting to escape from a pimp - she will avoid known shelters because the pimp might locate her and force her back into prostitution.

Many women in prostitution smoke cigarettes whether or not they abuse drugs. Others may be allergic to smoking and can't enter a facility where everyone in the program smokes. Addicted cigarette smokers who are not permitted to smoke in the shelter might use the addiction as an excuse not to enter a program.

The Physical Danger and Psychological Domination of Pimps are Barriers to Prostituting Women's Accessing Services

Pimps do whatever they can to prevent women from escaping them. Pimps actively prevent women from accessing alternative shelters. For example, pimps will only let women leave an apartment after 5:00 p.m. when they know most shelters or emergency services are closed. Pimps control the women's identification cards, so that women can not use them to escape.

Without ID, women can't buy a plane ticket, rent a hotel room, apply for welfare, rent a car, apply for an apartment, or even apply for a job outside prostitution. Without ID, she can't cash a check, open a bank account, make a bank withdrawal out of her own account, or even pick up cash that's been wired to her. Even if a family member wants to wire money to a daughter who wants to return home via bus or plane - when the girl doesn't have ID card to cash the check, she is unable to buy the ticket to return home.

Legal Barriers To Services For Women Escaping Prostitution

Pimps use sophisticated legal strategies to control women in prostitution, and to avoid getting arrested. One vicious tactic pimps use is to impregnate women under the guise of being "a family." When the woman attempts to escape and she has a child with the pimp/father, pimps then threaten her with kidnap charges and sometimes enlist the assistance of law enforcement. If she has a child with a pimp/father and she attempts to escape from a pimp by moving and not letting him know where she's living, he may sue for child visitation or custody, to force her and the children back into contact with him.

Pimps threaten to sue for sole custody if she says she no longer wants to prostitute for him or give him all of her money. She's in a catch-22 situation where the legal system forces her to contact him or she risks losing her child entirely. In addition, she might not be able to report a pimp's crimes without implicating herself as well, which again may put her at risk of losing her children. Pimps use the woman's own children to control women in prostitution, forcing them to earn financial quotas.

People mistakenly think a woman controlled by pimps can obtain a restraining order. While she can easily get a restraining order against him for herself, she can't prevent the pimp from visiting his own child. In order for the courts to issue a restraining order against a father from seeing his own child - there usually has to be physical harm to the child which demonstrates to the court that he's a danger to the child. During this time he is still allowed near the mother where he can threaten her, intimidate her, and sometimes physically coerce her into prostituting for him or giving him her money even if she has a restraining order against him for herself. Either the laws themselves or the interpretation of the laws must change if women are to be freed from this cruel legal bondage used by pimps.

Many people are in denial about the fact that parents pimp out their own children for prostitution and pornography. If a child tries to escape, she will be returned to the perp/parents by police. The only way not to be returned home is for a child to testify against her parents for their crimes -a very difficult thing for most children to do. If the child is listened to and believed, she is placed in a foster home. Even then, her parents can force visitation and force the return of the child to the home where the abuse and exploitation can continue - along with retribution for reporting them in the first place. There is little expert legal assistance for eleven year olds to escape pimp/parents under current laws.

These difficult legal scenarios cost money to fight, but women and children in prostitution do not have much money. Some attorneys have candidly reported that they fear pimps' or organized crime retribution if they take such cases. They may also fear becoming known in conservative legal communities as people who "defend whores." Free legal services are often offered by inexperienced attorneys who are not sufficiently educated in the complex legal issues faced by women and children escaping pimps.

Lack Of Transportation Is A Barrier To Escaping Prostitution

Other women, especially those escaping from hostage prostitution, may be in a town where they don't even have an idea where the nearest bus stop is - let alone how to reach where they'd have to go to receive services. It is a common tactic for pimps to take women into an unknown city or country so she does not how to get around and possibly seek help. Or he might take her to a location where she does not speak the language in order to prevent her escape. This makes it impossible for her to seek transportation and actually reach any services.

Even with a strong desire to get out of prostitution, without free and accessible 24-hour transportation to these programs, women won't be able to utilize the services that are being offered.

Lack Of Appropriate Clothing Is A Barrier To Escaping The Sex Industry

Women who are prostituting often do not have clothes appropriate for job-seeking. There is an urgent need for business attire for women to get out of prostitution. Many companies in Las Vegas will not permit new employees to begin job training unless they are dressed for the job. Job trainers have told women to leave and return dressed "properly" if they appear in the only clothing they own. The shame of this kind of encounter profoundly discourages women.

Fully aware that it limits the likelihood of escape, pimps do not allow women to have any clothing other than their prostitution costumes. Even if interviewing for a job that provides uniforms - the recruiter is unlikely to seriously consider a woman who interviews in a tube top, platform heels and a leather mini-skirt.

Barriers To Employment For Women Escaping Prostitution

It's difficult for anyone to get a job in Las Vegas, let alone someone escaping the streets. Even menial jobs require a health card, a sheriff's card, a work permit, a drug test and/or physical examination (often paid for by the job applicant). Today, some jobs even require a credit check. All of these take time to apply for and obtain, and there are fees involved with each. The offices where you obtain these cards are in different parts of town, so it can take a lot of time by public transportation. Some companies require the purchase of a uniform before she can report to work.

I once asked a manager at a large Las Vegas casino how much it would cost and what papers were required to get a job as a housekeeper at a starting salary of $9.00 per hour. Prior to her first day at work, he told me that she must have:

1. A work permit, also called a sheriff's card.

2. A health card, which means that she must be healthy enough to be eligible for group health insurance coverage. Women who weigh over 200 pounds are not hired for that reason alone.

3. A drug test.

4. Two sets of uniforms, special shoes, pantyhose, and possibly a hair makeover if their present style is too "street." Women have to buy all these clothing items, which they do not have in prostitution. All tattoos must be covered with band-aids supplied by the employee.

5. A telephone so that she can be called into work if they need her.

6. A babysitter or daycare if she has small children. Since most new employees in casinos are put on night shifts until they have proven themselves, this means finding night-time daycare for her children. If the casino has on-site day care for employees - she can't use it until she's been employed long enough to be off probation.

7. Transportation to and from work at odd hours because everyone has different shifts in the 24-hour casino work environment. In Las Vegas, aside from the Strip itself, the public transit system slows to a crawl on evenings and weekends. If her children are in day care or school, she must coordinate transportation with the hours of her shift which may change on a daily basis.

8. Must be able to speak and read English so she can understand schedule sheets and memos from the office, as well as instructions from her supervisor. Even to clean toilets, she must be literate and English-speaking.

9. Driver's license or state ID card in order to obtain the job and be able to cash paychecks.

10. An original social security card so the appropriate tax forms can be filled out.

11. A mailing address where the work application can be mailed to her since they are not offered on site. If she is homeless or frequently moving, she won't be able to obtain the cards she needs to start work. Casino jobs do not provide housing or offer discounts for onsite residence.

12. A GED (alternative high school diploma) or proof of enrollment in a program towards obtaining the GED.

13. She's required to join the union if there is one. This requires an application, fees, and possibly special classes, depending on the union.

14. She must be older than 18 to work in hotel house keeping. She must be at least 21 if any of her work is directly in the casino area. Access to temporary job agencies is limited since many women have criminal records including drug arrests, and many have been prostituting on the streets for many years. Any casino-related job - even a temporary one - requires the same work permits.

If she lies about her job history, that will usually be discovered by background checks and she will not given the job because she lied. If she's honest about her past, she may get a job but must wait to start until after a probationary period during which she demonstrates to the potential employer that she won't return to prostitution and/or other crimes. During this time, she has no income.

Unfortunately, there is an incentive to lie. One woman who wanted to escape prostitution had a massage therapist license from California where she had previously turned tricks in massage parlors. On one occasion in California she was arrested for solicitation although she was not convicted. The arrest motivated her to quit prostitution and move to Las Vegas, where she stayed with a friend while attempting to find legitimate massage work in Las Vegas. She applied for a job as a massage therapist at a physical rehabilitation center. Even though she was trying to start a new life by applying at a rehabilitation center rather than a massage parlor, the licensing board denied her application because her arrest had occurred within the past year. She was not allowed to work at the rehab center in any capacity until this probationary year had ended.

Language And Literacy Barriers To Services For Women Escaping Prostitution

Outreach must be conducted in first languages used by prostituting women. In the Las Vegas area, that means outreach should be conducted in Spanish, Korean, Chinese, Russian, Thai, and not just in English. The majority of staff in assistance programs for prostituted women speak only English although there are gradually growing numbers of Spanish-speaking staff – for example in the Las Vegas Metro Police Department, there are bilingual Spanish/English victim advocates.

If women do not speak English, even if they are able to break free and seek assistance, they have no way of finding out where to go, how to fill out forms, answer any questions, understand the rules of the shelter, or even be able to ask for help without the assistance of an interpreter.

If women are not U.S. residents or if they are undocumented, they might fear going to a shelter because it might lead to deportation. For many, being returned to their country of origin is worse than being controlled by pimps - so out of fear of deportation, they do not seek assistance in escaping prostitution. The following excerpt refers to Latinas seeking services, but much of it applies to women from any culture that is different from the shelter staff's own culture.

> Culturally appropriate services are especially important for Latina survivors of prostitution, trafficking and other sexual violence. Not only language, but country and regional nuances need to be addressed in order to meet the needs of Latin American victims of all forms of sexual exploitation. Limited English language skills restricts access to information about rights, services and options, thus increasing a feeling of dependency.

> A lack of translators, lack of bicultural/bilingual professionals, and lack of reading materials in the client's

native language all pose barriers for victims of sexual exploitation. At some battered women's shelters, for example, other Latina survivors or residents have been inappropriately asked to interpret, just because they happened to be in the agency at the time. The use of resident interpreters may cause embarrassment and silence when sexual violence is addressed. Rather than sharing personal or shameful information with shelter roommates or worse yet - bilingual child residents - a survivor may simply choose not to discuss her sexual exploitation.

While some Spanish language materials are better than none, the message is lost or distorted when dialect, differences in attitude/awareness of sexual exploitation and class differences are ignored. Materials offered to survivors must take into account: race discrimination, socioeconomic segregation, Spanish language limitations, and immigrant women's lack of knowledge about US laws. If a victim does not define her experience as abusive, no matter how adverse her experience, she will not seek help from violence prevention programs. Furthermore, the very label "victim" may exacerbate her feelings of shame and self-blame. Culturally sensitive screening that incorporates a range of references to sexual abuse can be helpful in reframing the abuse and shifting the responsibility to the perpetrators. The phrases me abusaron (they abused me), me falto el respeto (he disrespected me), me obligaron a salir con otros (they made me go out with others), are some of the many ways that Latinas may refer to sexual assault and sexual exploitation.[476]

Women in the sex industry, whether in strip clubs, escort agencies or street prostitution generally have poor literacy skills. Many also have learning disabilities. Many have poor eyesight but are without proper glasses and/or contact lens because of the expense and because they don't have health insurance. It is difficult to believe but there are women who earn thousands of dollars weekly in prostitution who can not read a restaurant menu and must point to the pictures to order a meal. This is a tremendous barrier to leaving the sex industry, because she can not read ads for assistance, fill out forms to enter programs, or fill out job applications, not to mention writing up a resume. Women who are illiterate experience great shame. They can't access Internet help or emergency services in a phone book. Often, women hide their inability to read and write because of a fear of appearing stupid.

Health Services Barriers for Women Escaping Prostitution

There is a lack of emergency services specifically for women leaving prostitution. Just as rape victims need medical care providers trained to meet their specific needs - so also do women escaping prostitution need specialized medical care. Women escaping prostitution are frequently in poor health. They may be malnourished, suffer from violence-related injuries, eating disorders, disability, mental illness, substance abuse, or involuntary drugging by pimps.

Since most women who are prostituting sleep during the day, they need nighttime services. Women usually decide to quit prostitution or seek emergency help during the hours that most social and medical agencies are closed.

Mental Health Services Barriers for Women Escaping Prostitution

Many women who enter prostitution have already been disabled by incest, rape, battering, or kidnapping. They may suffer from developmental disabilities after years of drug abuse which has caused chemical imbalances. These conditions make women vulnerable to prostitution in the first place. Once in prostitution she suffers various kinds of emotional damage from the stresses and traumas that are a part of prostitution. Common disorders are posttraumatic stress disorder, depression, drug and/or alcohol addiction, sexual compulsivity or sexual trauma reenactment, sexual dysfunction, eating disorders, schizophrenia, mood disorders, anxiety disorders, dissociative states, borderline personality disorder, maybe even multiple personality disorder. Women who entered prostitution at age 11-12 have not had a chance for normal emotional development, have not had a normal education or job training, and often do not even know how to act appropriately when they are not actively prostituting.

Many of these mental health problems can cause women to act in ways that are very challenging for service providers, who are usually not trained to calmly and respectfully connect with women who are angry, mistrustful, moody, high, paranoid, and manipulative. Some employees may assume women are being secretive or uncooperative when in fact they actually can't remember details in their lives. The workers may not be experienced enough to know how to tell the difference for example, between lying or being in a dissociative state. Some applications for assistance require information that the woman may not be able to remember because she's blocking out traumatic memories or because of memory damage.

An untrained service provider may assume that a woman seeking services likes prostitution, that she is protecting her captor/pimp, or that she is freely participating in criminal activities. It is essential for service providers to understand pimp brainwashing of women in prostitution. This sometimes takes the form of symptoms of Stockholm Syndrome in which people adapt to captivity in order to stay alive by bonding with their abusers. Like Patty Hearst, she may identify with her pimp's world view in order to stay alive. It may be more appropriate to schedule a medical appointment than to file a police report.

Lack Of Individual Health Insurance Is A Barrier To Escaping Prostitution

Few women getting out of prostitution have health insurance, and even if they're not in denial, they fear seeking help for emotional problems. Only one location in Las Vegas offers psychotherapy and psychiatric evaluations for women who have no health insurance but this agency is not easily accessible. It is unlikely that a stoned dissociative schizophrenic with no car and no money is going to be able to walk 15 miles across town to get help at the clinic. Even if she did actually locate this facility, she would probably not arrive during normal office hours.

Some women have health insurance through a nonprostitution job, or via Medicaid, or from a spouse's health insurance. Yet many are discouraged because of previous negative experiences. Others are afraid that if they seek help they might be locked up due to an outstanding warrant or behaviors they might confess to if they started talking about their problems. Most are afraid that their records could be used as evidence against them to prosecute them for crimes or to prevent them from getting future jobs and/or education.

These are not unreasonable fears. One man who had prostituted as a teen and who was photographed in pornography as an adult, decided to become a psychiatrist so that he could help people like himself. Following years of undergraduate work, he was admitted with a 4.0 grade average to a well-known university. He was suspended once the dean discovered his past history in prostitution.

Lack of Peer Support is a Barrier to Escaping Prostitution

Survivors of prostitution who staff an agency that offers services to women escaping prostitution have a unique ability to offer a special empathy and also to serve as role models for women who want to get out. Their very presence says, "If I did it, you can do it too."

Yet few survivors of the sex industry are open about their past - and for good reason. Formerly prostituted women who work in programs for women escaping prostitution – and who come out themselves – often suffer discrimination. Some years ago, I developed an alternative sentencing program in Los Angeles staffed entirely by ex-sex workers, most of whom had the appropriate degrees and experience, were out of prostitution and clean and sober for more than five years. Los Angeles County refused to fund the program unless non-survivors staffed it. The prejudicial attitude: "you can't trust them no matter how long they've been out of prostitution" reigned supreme in this flawed decision. When I told those who wanted to work with this program that the decision makers did not approve the program simply and solely because they had once been in prostitution many years ago - it was a devastating blow to all of them. Instead of feeling proud of their recovery and rewarded for their hard work to obtain degrees, they felt that the system was still punishing them whether or not they were in the sex business.

Lack Of Prostitution Awareness Programs In Secondary Schools Is A Barrier To Services For Women Escaping Prostitution

The average age a woman enters prostitution is between 12-14 years. Although schools now offer programs that teach children about drug abuse and contraception, there are few prostitution prevention programs in U.S. schools - and none in Las Vegas. We need to teach children the dangers of prostitution, common methods pimps use to lure them into hostage prostitution, as well as programs or alternatives they might access rather than turning to prostitution if they feel it's the only way to support themselves, or where they might go to find help should they need help in getting themselves, or someone they care about, out of prostitution.

Need For Coordinated Case Management To Help Women Escape Prostitution

When a woman is trying to leave the sex industry, she needs many things to assist her in making that major life change, often including medical care, housing, substance abuse treatment, a job, job training, legal assistance, psychological counseling, building up a credit report, parenting classes or counseling for children who may have been exposed to a culture of violence. Locating these services can be overwhelming, and sometimes it can't be done alone. Women escaping prostitution need an advocate and a coach - someone who can walk them through the maze, tell them where the services are located, encourage them to keep making small changes one step at a time, and to hold their hand while they walk through it so they don't give up.

While a woman might need a therapist to help her resolve emotional issues, she may also need someone to make sure she arrives at her appointments. Even if she has a medical doctor and a therapist, she may also need legal assistance to help her complete probation or to escape a husband who pimped her. Each of the various service providers may know only one part of her story, which makes coordinated care very challenging. There should be one advocate or case manager who is aware of the big picture of her entire life story and current needs.

Most programs that assist women getting out of prostitution emphasize education and job skill training as an alternative to prostitution. Either of these may take years, and costs may be out of her reach. Often the job or educational training comes with the same barriers as actually being hired for the job, such as her need for transportation, child care, literacy, appropriate clothing, empathic peer support, and financial support while she is pursuing alternatives to prostitution. To a woman trying to escape prostitution, it may feel like you're telling her that she has to climb Mt. Everest in order to get out of prostitution. She may have several children, no high school diploma (because she may have begun prostituting as young as 11), she may have learning disorders, emotional problems, or is in early recovery from substance abuse. It can be overwhelming to tell her that she can't support herself without prostitution until she obtains a GED.

You simply can't ask a woman to give up earning $200 a day to study for a GED for four years while working at a minimum wage job (if she's lucky) that might not even cover the rent for a seedy motel room. It's really not a reasonable alternative to prostitution. Instead, what seems to help many people over those years get out of the sex industry is to provide the women with jobs they can function in with their current skills - jobs that don't require more than a few hours of training, where her background is irrelevant to the employer, and where she receives immediate pay so that she doesn't have to turn tricks on the side to survive. The immediate pay boosts her self-confidence and gives her hope that she can stay out of prostitution. Later, she can take evening or online courses to get a degree of some kind.

13

Attitudes toward Prostitution and Sexually Coercive Behaviors of Young Men at the University of Nevada at Reno

Melissa Farley, Mary Stewart, and Kyle Smith

We investigated the impact of Nevada legal prostitution on the attitudes and behaviors of college-aged men toward women. The legal sexualized objectification of women in brothels is likely to have an impact on the community, especially on men's relationships with and attitudes toward women. The objectification of women and of sex itself is integral to many men's definitions of masculinity. Nonetheless, these beliefs impair men's ability to relate to women as equals, and interfere with their ability to form and maintain intimate relationships with women.[477]

There is much psychological research demonstrating adverse effects of pornography specifically, and television generally on men's attitudes and behaviors toward women.[478]   Therefore we predicted that legalized prostitution, the nearby presence of prostituted women in legal brothels who might well be seen by men as 'live pornography', and a state-wide culture of prostitution would have a powerful impact on men's attitudes and behaviors toward women in general.

In 2002 Ann Cotton, Melissa Farley and Robert Baron published a study of United States college students' attitudes about prostitution.[479] We asked about prostitution myth acceptance and rape myth acceptance in a sample of 783 university undergraduates from California, Iowa, Oregon and Texas. Since others had described an association between violent behaviors against women and culturally supported attitudes that encourage men to feel entitled to sexual access to women, to feel superior to women, or to feel that they have license as sexual aggressors,[480] we wanted to investigate this relationship with respect to prostitution.

Rape myths[481] and prostitution myths are components of these culturally supported attitudes that normalize violence against women.[482] Rape myth acceptance is related to sexual aggression.[483] Prostitution myths justify the existence of prostitution, promote misinformation about prostitution, and contribute to a social climate that exploits and harms not only prostituted women but all women. In this 2002 research, we found that among college students, acceptance of rape myths was significantly positively correlated with acceptance of prostitution myths.[484]

Further analyzing the data obtained from the college student sample, Schmidt, Cotton and Farley[485] found that men who reported sexually violent behavior against their non-prostitute partners endorsed significantly more prostitution myths than men who did not report sexual violence. In addition, those men who had bought women for sex (johns) reported significantly more sexually violent behaviors against their nonprostitute partners than those men who had not bought women for sex (non-johns).[5]

Cotton, Farley and Schmidt then compared college students who said that they had engaged in sexual coercion with those who were not sexually coercive. The sexually coercive men were significantly more accepting of prostitution myths than those men who did not report having been sexually coercive with their partners.[486]

In 2006, Dr. Mary Stewart, her student Kyle Smith and Dr. Melissa Farley used the same questionnaires to compare young men who were students at University of Nevada at Reno to the previously studied U.S. college students, those in the Cotton, Schmidt and Farley studies.

The University of Nevada, Reno (UN-R) is located in a medium sized metropolitan area in northern Nevada. Eighty percent of the 11,500 undergraduate students at UN-R, are Nevada high school graduates. Most of these students grew up in northern Nevada rather than in the larger metropolitan area, Las Vegas, to the south.[487] We felt that the Reno university students were noteworthy in that these young men were attending a school that was not only physically closer to a legal brothel[488] than any other college or university in the United States but was a school which existed in a cultural milieu that defined women as objects for sale.

Although the university is located in Washoe county where prostitution is illegal, nearby counties Lyon, Storey, and Churchill have legalized prostitution. Pimp Joe Conforte's brothel was 20 miles from the

University until it was locked down after Conforte fled to Brazil in 1991 subsequent to his arrest for income tax evasion.[489] Additional charges filed against Conforte in 1998 include bankruptcy fraud, aiding and abetting, money laundering, witness tampering, racketeering, conspiracy and forfeiture. Conforte's priors include at least 2 convictions for which he served time: in 1959 for trying to extort Washoe County District Attorney Bill Raggio, and in 1963 for income tax evasion.[490]

Conforte's nephew, a Hell's Angels chapter vice-president, still controls a brothel that is also close to the University of Nevada. Conforte himself is rumored to be homesick for Nevada and is rumored to have snuck in and out of the U.S. on several occasions. He seems to take pleasure in telephoning competitor pimps from Brazil and informing them that he is watching their businesses. He is reportedly heavily involved in Brazilian prostitution.

Other pimps in the Reno area include Dennis Hof, an enterprising pornographer. Hof is one of several legal Nevada pimps who pose as community benefactors, giving away turkeys at Thanksgiving, or offering Iraq war veterans a golfing weekend that includes free prostitution.

The students who were included in this part of the study had grown up in a climate in which legal prostitution was an ordinary part of the culture. Brothel visitors could easily obtain directions and transportation to the legal brothels, as well as being able to buy women in illegal escort prostitution in the local casinos. Local pimps transport their stables of prostituted women to Reno-area nightclubs, as advertising for prostitution. Prostitution in Nevada is accompanied by a casino and entertainment industry that highlights women as sexually available - taxis and buses advertise nude reviews and "gentleman's clubs" and neon signs and giant billboards present cars, alcohol, gambling and women as essential components of an evening of entertainment. Although the latter is present in many US cities, legal prostitution is not. Prostitution was woven into the cultural fabric of northern Nevada.

We wondered if a cultural environment in which women were legally for sale in prostitution would affect the college students' attitudes toward prostitution, toward rape, and their attitudes toward women in general. We expected that these young men would be more inclined than other college students to normalize prostitution and the attitudes toward women that prostitution engenders, since the state of Nevada legally sponsors prostitution. We obtained approval from the University's Institutional Review Board to administer an anonymous questionnaire to young men at University of Nevada at Reno. In 2006, we obtained questionnaires from 131 young men who were attending undergraduate classes at UN-R.

Psychologist Ann Cotton worked with us to compare data from the Reno men with men from universities in California, Texas, Oregon and Iowa via statistical analyses. We ran a one-way between-groups multivariate analysis of variance (MANOVA) that compared the means of groups (UN-R men vs. other men college students) on a number of different but related variables.[491]

We found significant differences between the attitudes of the Reno students toward prostitution and those of the other college students.[492] The Reno students significantly more often endorsed beliefs that prostitution is a normal activity which should be mainstreamed, such as "There is nothing wrong with prostitution,"[493] "Prostitution should be treated no differently than any other business,"[494] "Prostitution should be legalized," or "decriminalized,"[495] "There is nothing wrong with having sex for money,"[496] "Arresting men who patronize prostitutes causes more problems than it solves,"[497] "It's OK for a man to go to a call girl if his wife doesn't find out,"[498] and finally, "I would use an escort service or patronize a call girl if I knew it was safe to do so."[499]

The young men from Reno significantly more often than the other college students subscribed to the unproved and mistaken 'catharsis' theory of prostitution: they assumed that the very existence of prostitution served as a crude pressure release which somehow decreased the likelihood that other (nonprostitute) women would be raped.[500] When compared to the non-Reno students, the Reno young men significantly more often said that they also liked nonrelational sex, that is, sex with no emotional involvement.[501]   A hallmark of the sex of prostitution is its nonrelational, no-emotional-connections nature.[502]

In spite of these extensive rationalizations for the existence and normalization of prostitution, the young men from Reno also significantly more often than the non-Reno college students believed that prostitution lowers the moral standards of a community.[503] Further research into this seemingly contradictory set of opinions is needed, although it should be noted that in a separate series of cross-cultural interviews with men who frequently buy women in prostitution, we noted a similar set of opposing attitudes.   On the one hand, they defended prostitution as a socially necessary institution and on the other hand, they felt nagging guilt and self-contempt for their actions in buying women.[504]

The young men from Reno were more likely than men who attended universities farther away from legal brothels to believe the myth that prostitution sex would turn them into better lovers.[505]   They were also significantly more likely to believe that women became prostitutes because they liked sex,[506] and that prostitution was a choice that women should have.[507] These differences are statistically robust, suggesting that there are major differences in how the Reno and the non-Reno students viewed prostitution.

The Reno students more strongly endorsed items that reflected support for prostitution as a reasonable option for the next generation. For example, they felt that it would be acceptable if their sons went to prostitutes[508] and that it would be acceptable if their sons went to brothels.[509] The young men from the University of Nevada-Reno also significantly more often than young men from other U.S. universities, felt that it would be acceptable if their daughters grew up to be prostitutes.[510]

Compared to non-Nevadans, the Nevada university students significantly more often endorsed statements that justified rape and sexual violence against prostitutes. The Reno men significantly more often felt it was "ridiculous for a call girl to claim she's been raped by a customer,"[511] and more often stated that "if a man pays for sex, the woman should do whatever he wants.[512] These deeply sexist attitudes toward women in prostitution - the notion that they exist simply for men to use them - are core beliefs that justify any abuse, especially rape, of women in prostitution who are assumed to be worth less than other human beings. The Reno men endorsed the statement that "prostitution is an exploitation of women's sexuality" significantly *less often* than the non-Reno men.[513]

We also used a MANOVA statistical analysis to investigate the differences in rape myth acceptance between the Reno college men with men from universities in California, Texas, Oregon and Iowa. Rape myths are culturally supported attitudes that normalize rape. Over the years, psychologists have found that acceptance of rape myths are associated with sexually aggressive behaviors.

Where there were significant differences between the Nevada and non-Nevada students, the Nevada students more strongly endorsed rape myths.[514] For example, the Reno men significantly more often endorsed the rape myth that "Women generally find being physically forced into sex a real 'turn-on.'" [515] The Reno men normalized sexual coercion in that they significantly more often endorsed the statement "If a woman is willing to "make out" with a guy, then it's no big deal if he goes a little further and has sex." [516] The Reno men significantly more frequently than the non-Reno men erroneously assumed a class difference in rape: "Men from nice middle class homes almost never rape."[517]

It was noteworthy that the Reno college students reported that in the year before they filled out our questionnaire, they used prostitutes significantly more often than the non-Reno college students.[518] The University of Nevada-Reno students significantly more often went to strip clubs[519] and to massage parlor brothels[520] than those students who were from universities elsewhere in the United States.

There were also differences in the two groups of men in their pornography use. The young men from Reno watched significantly more video pornography[521] and viewed more Internet pornography[522] than the non-Reno college students. There was an eight-year difference between the time when the non-Reno university students and the Reno university students responded to our questionnaires. It is possible that the statistically significant differences in the use of video and Internet pornography is at least partly a result of the passage of time during which there is the increasingly widespread use of Internet pornography by all young men, not only those men attending a university that is close to a legal brothel. Nonetheless, the difference between these two groups of men is in the predicted direction.

In conclusion, the cultural mainstreaming of prostitution in Nevada had a significant effect on the attitudes toward women and the behaviors of young men attending University of Nevada, Reno. In contrast to students from other parts of the United States where prostitution is illegal, the UN-R students were significantly more accepting of prostitution, more accepting of several kinds of sexual violence against not only women in prostitution but also nonprostituting women.

This suggests the powerful influence of a culture that offers women for sale. These findings reflect the sexual objectification and sexual violence against women occurring in a cultural environment where women are legal objects for sale. Along with their assumption of prostitution as normal, the Reno students held a number of sexist stereotypes about women in general.

The intrusion of legal prostitution in Nevada's culture affects men's level of acceptance of the institution of prostitution itself. They were more accepting of prostitution and of the nonrelational sexuality of prostitution than were college students in other locations. They justified the existence of prostitution by subscribing to the myth that if men go to prostitutes they are less likely to rape women.

In fact, Nevada's rate of rape in 2004 was 40.9 per 100,000 population - higher than the U.S. average (32.2) and was significantly higher than rates of rape in California (26.8), New York (18.8) and New Jersey (15.3). Las Vegas (44.7) and Reno (41.3) rape rates were significantly higher than rape rates in other major U.S. tourist destinations such as Los Angeles (23.2), and San Francisco (24.5).[523]

The Reno university students endorsed several rape myths that justified sexual violence. They were sex industry consumers. They were significantly more likely than the non-Reno students to use prostitutes, to go to strip clubs and massage parlors, and to use both video and Internet pornography.

The Reno university students normalized prostitution for their sons and daughters as well as themselves. They considered it acceptable for their future sons to use prostitutes and for their future daughters to become prostitutes.

The Reno students failed to see prostitution as sexual exploitation, while at the same time justifying acts of sexual violence against women in prostitution. They assumed that it would not be possible to, for example, rape a call girl. This attitude is typical of men who buy women in prostitution, and it places prostituted women in harm's way, drastically increasing the probability that johns will rape them.[524]

14
Adverse Effects of a Prostitution Culture on Nonprostituting Women

Make no mistake: prostitution laws are not only about prostitutes. They keep all women under control. At any time, any woman can be called a whore and treated like one. - Claude Jaget [525]

Prostitution not only harms the women in it, it also promotes sexist attitudes and sexually aggressive male behavior toward all women in the community. Prostitution in Las Vegas and in the state has adversely affected all women. Mainstreaming prostitution takes both public and private space away from women, in that all women are treated as prostitutes. Assuming the right to treat women as prostitutes means that they are treated as if they are not human, thus harming both prostitute and nonprostitute women.

What is it like to live in a working class, non-gated community in Las Vegas? In 2006, a Tropicana Avenue resident said that businesses and residents alike are afraid of the war zone atmosphere generated by the pimps: "My roommate is a female. She can't get out of her car without someone thinking she's a prostitute… it's a total kill zone of drugs and prostitutes…It's really scary."[526] Sexual harassment becomes more common in areas with a high density of sex businesses.

A Thai woman visited Las Vegas and was constantly harassed by pimps and tourists who assumed that she was sexually available or prostituting. With a context of racist portrayals of Asian women as subservient sex objects advertised on the Strip and also in popular books like "Memoirs of a Geisha," Pueng Vongs walked down the Strip and "was stopped twice on a recent afternoon, each time by Anglo-looking men who asked similar questions. They wanted to know where I was from, what I was doing there, how long I was staying. They stared at me like a plump, glistening prime rib roast centerpiece at a nearby buffet."[527]

A woman stated that "There's this whole generation of girls growing up in Las Vegas, surrounded by the industry and they naturally fall into it. It's a normal everyday thing."[528]   Another woman reported that her 18 year-old daughter served soft drinks in Las Vegas clubs, but that it appeared that she was being groomed by club owners for prostitution. Watching out for her daughter's safety in Las Vegas, she said, felt like "standing in the middle of a river, trying to hold it back."[529]

In the United States, girls are sexualized at a young age. The American Psychological Association reported on this phenomenon in 2007. The APA Report noted that every media form provided evidence of the sexualization of women, including television, music videos, music lyrics, movies, magazines, sports media, video games, the Internet, and advertising.[530]   Girls as young as 4-8 years of age are targeted and sexualized.

While 'ho and heroin chic clothes for women have been in the mainstream for 15 years, today girls are the market and prostitution is specifically being glamorized for their age groups. Bratz dolls, marketed to 4-5 year olds, are dressed in stripper outfits with miniskirts, fishnet stockings and feather boas. Thongs, and other stripper clothing, including underpants labeled 'hottie' or 'wink, wink' are designed for and marketed to young girls. Girls' Halloween costumes are eroticized.[531]   Prostitution is marketed as a fun job to young children.

The prostitution of young women's sexuality is exemplified in the "Girls Gone Wild" video series, where   sexual boundaries have mostly disappeared,   and   Youtube   and   MySpace   promote   the   widespread dissemination of private sexual thoughts and photographs. "Whether it's 13-year-olds watching a Britney Spears video, 16-year-olds getting their pubic hair waxed to emulate porn stars or 17-year-olds viewing videos of celebrities performing the most intimate acts, youth culture is soaked in sexuality."[532]

Hip hop, with its virulently racist and misogynist portrayals of young women, is a major influence on youth in Las Vegas. Women in hip hop are defined as 'hos and bitches, treated with contempt and often violence.[533]   Targeting young African American women with a brutal contempt, much of hip hop teaches young men that the value of women, especially Black women, is between their legs.[534]   Pimping has been glamorized for boys, by for example the cartoon 'Lil Pimp, in which a white 9 year old boy runs a stable of 'hos with his pimping pals.[535]

Even when women protest sexual harassment in Las Vegas, described by reporter John Smith as "the city that has raised sexual exploitation to a billion-dollar legal racket,"[536] the legal system doesn't really take their claims seriously. For example, a Riviera casino employee was sexually assaulted by a show producer in front of several witnesses, including the hotel president and another official (who herself acknowledged experiencing similar abuse from the same producer). Yet the producer was not fired and was permitted to plead to a minimal charge that amounted to a slap on the wrist.[537]

What happens to nonprostituting women employees in clubs like Hooters or Las Vegas casinos? According to a defense lawyer, women whose jobs require them to wear costumes accentuating their breasts and exposing their legs should reasonably expect sexual harassment?[538] In Las Vegas casinos, many cocktail servers are required to dress to look like they are prostituting. As a personnel director of one casino stated in a letter to the Equal Opportunity Employment Commission, "The scantily clad female is a symbol of all that is popular and expected in Las Vegas."[539]

The "fantasy" prostitute sold everywhere in Las Vegas can result in very real sexual harassment for waitresses. The job of serving cocktails is highly sexualized in Las Vegas, with even the name of the job sexualized as "bevertainer."[540] Sexual harassment of cocktail servers by drunk customers who spend large amounts of money is tolerated by managers of strip clubs and casinos. These behaviors include verbal sexual harassment, ogling, grabbing, pinching, prodding, among others.[541]

There is a blurring of the boundary between prostitution and restaurant serving in Las Vegas. For example waitresses at a club in the MGM Grand are expected to offer tequila shots from their bare breasts and elsewhere on the Strip waitresses are expected to dance with customers.[542] Women dealers at the Rio casino are required to wear thong bikinis at night.[543]

Not surprisingly, the Las Vegas Rape Crisis Center reported that a number of their clients are cocktail waitresses who have been raped. One such case involved a cocktail waitress at the Bellagio who was pressured by a hotel manager to spend some time with a gambler who had lost a great deal of money.[544] He wanted prostitution and she refused. In order to keep her job, the woman eventually agreed to a drink in Light, the Bellagio club. In public at the club, the man raped her with his hand. Bleeding, she asked for help but the hotel management did not call an ambulance. Instead, she drove herself to the hospital and later attempted to file rape charges against the customer. The Clark County District Attorney refused the case. This particular case was cited by at least one law enforcement source as an instance of the Las Vegas problem of "public corruption." It is also emblematic of community standards in a context where sex is a job for women.

Women blackjack dealers at casinos in Las Vegas generally don't dress like cocktail servers, yet they are still subject to sexual harassment and abuse from customers who are drunk and losing money. Customers have been known to make racist remarks to women blackjack dealers, and have threatened to rape or shoot them.[545] The men dealers however, do not encounter this kind of abuse. The women dealers have an especially difficult time controlling angry card players who stereotype women as sexual and servile rather than competent and in-charge.[546]

I asked Kelly Langdon, the Nevada State Rape Prevention Coordinator, how she understood the relationship between legal prostitution and the high rates of rape in the urban areas of Nevada. Emphasizing the inequality of the prostitution transaction, whether legal or not, she said that legal prostitution "creates an atmosphere in the state in which women are not seen as equal to men, are disrespected by men, and which then sets the stage for increased violence against women." [547]

A Las Vegas rape crisis counselor spoke bluntly about the relationship between the sex industry and the city's high rape rates. " Men think they can get away with rape here," she told me.

Data from the 2004 FBI Uniform Crime report validates these analyses and raises the possibility of an association between legalized prostitution, the state's prostitution culture, and rape rates in Nevada. The Nevada rate of rape was higher than the US average and was twice as high as New York's rate of rape. The rate of rape in Las Vegas was three times greater than that in New York City.[548]

Table 7. U.S. Rape Rate per 100,000 Population by *State*

Nevada	40.9
U.S. average	32.2
California	26.8
New York	18.8
New Jersey	15.3

Table 8. U.S. Rape Rate per 100,000 Population by *City*

Las Vegas	44.7
Sparks-Reno	41.3
San Francisco	24.5
Los Angeles	23.2
New York	14.0

Australian legal prostitution also seems to have an effect on sexual violence. Contrary to popular belief in that country, legal prostitution does not prevent rape and sexual assault. Within the state of Victoria, where prostitution is legal, overall crime rates decreased, while rape on the other hand, increased by 11%.[549]

Because of the mainstreaming of prostitution in U.S. popular culture, other kinds of sexual exploitation and violence are also trivialized or ignored. The complex issues causing women to enter prostitution, the role of men who buy women, and the suffering of women in prostitution are simply not understood by many people. People ignore what is in plain sight, or pretend it isn't as bad as it really seems.

Most researchers don't include cartoons about the topics they are studying. But I think that cartoons are a valuable means of communicating complex information. The following 7 cartoons seemed like an alternative and light-hearted way to address these deadly serious issues. I was fortunate to meet bulbul, a feminist artist who has drawn cartoons about the Vietnam war, the women's movement and the labor movement.[550] I asked her if she would be willing to draw some cartoons illustrating the paradoxes in how people in the U.S. think about prostitution and trafficking. She told me, "well, they may not be very funny, but they will be a visual." Feminist comedian Betsy Salkind made major contributions to the cartoons. The young girl who says, "My dad's my biggest client," the john who says, "I hate it when they act like I'm raping them," and the customer who says, "You have breasts, don't you?" - were all written by Betsy Salkind. The lines that are not so funny were written by me. Jody Williams contributed the lines to the cartoon contrasting legal and illegal prostitution. Bulbul's drawings, and her ability to eloquently express complex emotions with a few strokes of the pen never fail to amaze and delight me.

15
Conclusion: Legalization of Prostitution, a Failed Social Experiment

At the end of a long day in court in 2007, Las Vegas Prosecutor Mary Brown asked, "What the hell kind of world do we live in?" She had spent the day in a special Las Vegas court with children who had been charged with the crimes of solicitation, loitering for prostitution or being a minor in a gaming establishment. The rapid expansion of the prostitution of children is one of the known consequences of a sex industry the size of that in Las Vegas.[551] Prostitution and trafficking of 11 to 17 year old girls into Las Vegas is rampant[552] and should be considered a social and political crisis in the city.

Prosecutor Brown lives in a world - Las Vegas - where trafficking of women and girls into illegal prostitution is part of a multibillion-dollar prostitution industry. In Nevada, legal and illegal prostitution are connected, with the state's legal prostitution providing a cultural cover for the state's illegal prostitution.

Nevada is a world where judges' re-election campaigns may be paid for by sex industry racketeers, and where politicians' zoning decisions may be bought and paid for by organized criminals. It's a world where the older organized crime groups like La Cosa Nostra have been moved into second place in the sex industry by the newer and more violent internationally organized crime groups. It's a world where revenue from prostitution generated by international criminal networks has been connected with weapons trafficking.

It's a world, according to Kathryn Farr, in which the criminal networks that control sex businesses are inseparable from those she calls "corrupt guardians."[553] In other words, an illegal sex industry the size of that in Las Vegas, generating $1-6 billion a year, does not exist solely because of highly organized crime groups. It exists because import-exporters and other businessmen, restauranteurs, travel agents, gamblers, community leaders as well as police and politicians are involved to a greater or lesser degree. And almost everyone knows about it.

We live in a world where the "disrespectful, degrading, contemptuous treatment of women is so pervasive and so mainstream that it has just about lost its ability to shock."[554] It's a world saturated with pornography that humiliates and injures women. The world wide web's most frequent use is accessing pornography, which is indistinguishable from prostitution because *it is* photographs of prostitution. Videos that mainstream prostitution-like activities such as "Girls Gone Wild," generate billions of dollars. A world where Clinique moisturizer, using the pornographic gaze, advertises a woman with lotion spattered like semen across her face to simulate the john's triumphant, humiliating ejaculation into her eyes. [555]

Cultural mainstreaming of prostitution is at pandemic level. The website seekingarrangement.com posts photos of rich men who seek a relationship which is "usually between an older and wealthy individual who gives a young person expensive gifts or financial assistance in return for friendship, intimacy or sex." "Welcome to 2007. This is how life is. You always want to upgrade to the bigger and better thing," said a woman seeking a regular john who would pay her a monthly salary in exchange for sex."[556] A euphemism used for prostitution in Korea is "compensated dating." In Nevada, the parallel term is "mutually beneficial arrangement."

We live in a world where girls and women are sexualized for men's use. The United States is a place where younger and younger girls are used for sex, not only by pedophiles but by men who have been culturally conditioned to sexual arousal when they see children. As a 2007 report by the American Psychological Association warned, as girls become more sexualized, a market for sex with children is cultivated.[557] The massive expansion of online pornography, with an increasing emphasis on sexualized images of children and "amateurs," when combined with children's extensive use of the Internet, together expand pedophiles' niche in the sex industry, and has led to the sexual abuse of children.[558]

We live in a world where sexual assaults of children (also called sexual abuse, child prostitution, and incest)[559] are at pandemic level but there seems to be no political will to address these crimes against children. Once having been sexually assaulted in childhood, girls are 28 times more likely to grow up into prostitution than non-sexually assaulted children.[560]

A 2007 Las Vegas escort prostitution advertisement promoted "Slave girl direct to you." We live in a world where boys and young men are swamped with messages that normalize sexual aggression. A school psychologist described sexual aggression as having "permeated our society,"[561] with acts of sexual aggression committed by boys under age 10 becoming increasingly common.

We compared young men at University of Nevada, Reno with other US college students and found differences in their attitudes toward prostitution, their acceptance of rape myths and other sexually exploitive attitudes toward women. As discussed in Chapter 13, the prostitution culture in the state of Nevada affects young men's attitudes and behaviors toward women. More often than college students outside of Nevada, young men studying in Reno saw prostitution as normal, assumed that it was not possible to rape a call girl, and were more likely to use women not only in legal prostitution but also in illegal prostitution.

We live in a world of sex, race, and class inequality in which middle school students at a career day in California were told in 2005 that stripping and exotic dancing were excellent careers for girls. A man who spoke about potential jobs told a group of students that strippers can earn very good salaries, especially if they have breast enlargement surgery. "For every two inches up there, it's another $50, 000," he enthusiastically told the girls.[562] It's a world where a Las Vegas private school's coach for cheerleaders was arrested for pimping women in escort prostitution.[563]

Prostitution is a business rooted in social inequality: the inequality between men and women, between the rich and the poor, and between ethnic majorities and minorities.[564] Legal prostitution has set the stage for discrimination against women, especially those who are most vulnerable: poor and ethnically marginalized women.

We live in a world where discrimination against poor women is institutionalized as "sex-positive," a world where in 2005 a group of women who promote prostitution as a reasonable job for poor women received more funding from Canadian government and nongovernmental sources per year than any other women's groups in Quebec. The prostitution advocacy group Stella had an annual 2005 budget of $500,000 (Canadian dollars) while the Coalition of Rape Crisis Centres with 29 members in 17 regions of the province of Quebec had an annual budget of $420,000 (Canadian dollars).[565]

The cruelty of the state against some of its most vulnerable citizens is reflected in Nevada's system of legal prostitution. The state's archaic decision to offer up women for men's legal sexual use is a failed social experiment.[566] It has not increased prostituted women's psychological or physical safety. Legalization has not decreased women's economic exploitation by pimps or violence against them by johns or the psychological trauma resulting from these harms. It has not reduced the social stigma of prostitution. As Lenore Kuo wrote: "the treatment of prostitutes under legalization is horrific; women are stripped of some of their most fundamental rights, dignity, and humanity."[567]

Legal prostitution in Nevada and its illegal offshoots provide a welcoming environment for local pimps and also for international pimps and johns. Both legal and illegal prostitution support organized crime, which in turn corrupts the political and judicial system. In the county nearest to Las Vegas with legal prostitution, County Commissioner Trummell had a close encounter with the material reality of pimping. In 2006 Trummell cooperated with the FBI in surreptitiously recording a legal pimp's bribe for her favorable vote regarding the location of his brothel.[568]

Legal prostitution increases sex trafficking because of the market for sex, that is, because of men's demand for 24/7 purchased sexual access to women and girls. It's actually deceptive to make a distinction between trafficking and prostitution because the implication is that it is the *distance she is moved* in order to be sold for sex that matters rather than being sold, used, and prostituted per se. The distance is irrelevant. What's relevant is what is done to her in prostitution, the sale of and sexual use of a human being.

The authors of a study of legalized prostitution in Europe concluded: "Legal control of prostitution targets the outward appearance rather than the conditions in which women find themselves."[569] That observation holds equally true for legal prostitution in Nevada. Governments with legal prostitution seem to be more concerned with public order than with abuse and violence.

Marjan Wijers and Lin Lap-Chew documented abusive conditions in European prostitution that were identical to what we and others have documented in Nevada: psychological and verbal abuse, physical abuse, coercion into performing specific sex acts, confinement, police harassment and arrest, sexual assault, no right to refuse a customer or a mandatory minimum number of customers per day, isolation, denial of days off, imposition of 12 or more hours of work daily, no or extremely limited access to medical care, confiscation of personal belongings, denial of right to use condoms, withholding of pay, no proper sleeping accommodations, compulsory AIDS testing, and inadequate food.[570]

In the Netherlands, politicians who previously supported legal prostitution have changed their minds about it. "I have often doubted since we legalized the brothels, whether we did the right thing," said Femke Halsema, a member of the Dutch Parliament who advocated legalization. "For me, it was a question of emancipation and liberation for the women. *But for now it is working the other way.*"[571]

Trafficked women in Nevada, as elsewhere in the United States, are sometimes viewed as illegal immigrants. It is a miscarriage of justice when women and girls in prostitution are treated as if they were criminals rather than victims. In Las Vegas, women in prostitution are more harshly treated than the predators who buy them. Johns in Las Vegas on the other hand, especially those who buy women in indoor prostitution, are unlikely to be arrested.

Police and independent researchers in the Netherlands, where prostitution is legal, have shown that traffickers use work permits to bring foreign women into the Dutch prostitution industry, masking the fact that women have been trafficked, by coaching them to describe themselves as independent "migrant sex workers."[572]

Similarly, 30-day work permits have been used to traffic women into Nevada's legal and illegal sex industry. Traffickers and pimps coach women to dismissively smile and say they're happy when approached by law enforcement or service providers. Today, trafficked women comprise 70-80% of those in the windows of Amsterdam, and in Dutch legal brothels elsewhere in the country.[573]

Some subscribe to the myth that legal brothel prostitution keeps economically strapped rural counties afloat. Others have challenged that assumption with evidence from county budgets. Nye County Commissioner Trummell said, "Many people believe the brothels contribute significantly to our economy, but they don't. We have six brothels and we receive a total of $180,000 in fees."[574] Trummell estimated that the total 2005 brothel revenue was less than 1% of the entire county budget.

Similar observations have been made in other Nevada counties with legal prostitution. According to Churchill county newspaper editor Josh Johnson: the county Comptroller found that in 2004, "the cost of related law enforcement expenses exceeded revenue obtained from prostitution. *The county was, in fact, subsidizing the industry…*"[575]

Furthermore, the income generated from fees and licenses in Nye County did not appear to cover the expenses of regulating the brothels. Nye County Sheriff Anthony De Meo said he needed 30 more deputies in 2005 just to keep up with the crime in the county and in order to sufficiently regulate the brothels. De Meo said he would need to hire an additional 2 full time investigators as well.[576]

While often complicit in keeping the issue of prostitution off the radar screen of most people, nonetheless Nevada politicians and urban developers rarely publicly defend prostitution. Former Nevada Gov. Kenny Guinn stated, "I'd prefer not to have it at all in the state of Nevada."[577] In 2007 both Democrat (Hillary Clinton) and Republican (Mitt Romneypresidential candidates spoke out against legalized prostitution.[578]

Las Vegas casino developer Steve Wynn has spoken out against legal prostitution:

> We have outgrown legalized prostitution. The existence of brothels in Nevada is just one more item that out-of-state media people use to denigrate the quality of life in Nevada...It is not good for Nevada's image to have ...legalized 'cat houses' and the sooner we put that image behind us, the better we will be. [579]

Las Vegas Mayor Oscar Goodman and the state's brothel lobbyist are among the very few enthusiastic proponents of prostitution. Nevada Brothel Association lobbyist George Flint said, "I wish Las Vegas would just get smart and legalize prostitution, but due to the Mormon influence I don't think that will happen."[580] Joining the Nevada Brothel Association is Las Vegas Mayor Oscar Goodman who has proposed legalizing prostitution in Las Vegas as a "redevelopment tool"[581] In other conversations, Goodman has talked up lap dance prostitution and escort prostitution [582] and has suggested turning "old motels into beautiful brothels."[583] He has been described as "a lawyer for the mob."[584] For 35 years, Oscar Goodman was a defense attorney for gangsters and members of organized criminal groups. In *Of Rats and Men: Oscar Goodman's Life from Mob Mouthpiece to Mayor of Las Vegas*, John Smith describes Goodman's clients who included Meyer Lansky, Nick Civella, Anthony Spilotro, Frank Rosenthal, Nicky Scarfo, and Vinny Ferrara, among many others. Though no other connection between Goodman and the Mafia has been proved, he has often been accused of being more than just a mouthpiece for organized crime.[585]

In 2003, a Magellan Research poll showed that a majority of Las Vegans opposed Goodman's idea. Jodi Tyson, former director of the Nevada Coalition Against Sexual Violence, thinks that legalizing prostitution essentially legitimizes abuse of women. Making it legal simply moves responsibility for the problem from street thugs to the government, she said, "and then we become the pimps."[586] Mary Sullivan makes a similar point regarding profits generated by Australian legalized prostitution. She points out that legalization has effectively commodified women in prostitution, to the great benefit of corporations that own sex businesses and financial institutions which profit from men's buying women for sex.[587]

Women in the Nevada legal brothels are not contented women earning huge amounts of money. They describe their living conditions more often than not as jails in which they live under virtual captivity. As a consequence of servicing hundreds of johns in prostitution, the women who have prostituted for years in the Nevada legal brothels exhibit signs of depression and psychological dissociation, and sometimes symptoms of long-term institutionalization.

Ultimately, governments are responsible for promoting sex and race equality and for eliminating poverty. When governments fail at these tasks, women's choices become more and more limited. Prostitution and trafficking are the direct consequences of these governmental failures.

We live in a world where women are increasingly channeled into prostitution as their opportunities for work in other sectors of the economy shrink. A Yemeni woman who prostituted, tearfully accused her government of making her "worthless and of no value, oppressing us with these unstable conditions, moreover forcing us to indulge in actions that will haunt us for generations to come."[588] The prostitution of desperately poor women in Yemen may seem worlds apart from the prostitution of women and girls in the United States. But it is closer to home than you might imagine. Globalized economies that feminize poverty while enriching male-dominated economic systems ultimately harm all of us.

Will Nevada, like Australia, soon rank 'sexual services' (prostitution) higher and higher among service industries' revenue? Brothels and other sex businesses are planned for Australian business centers and shopping malls. An Australian economic analyst predicted that the sex industries' revenues will increase by more than $2 billion (US dollars) by 2010. This increase in revenue from the business of sexual exploitation is equivalent to a 6.8% annual increase at a time when the overall Australian economy is only growing at a rate of 3%. [589]

Dr. Howard Meadow directs a diversion program at the Las Vegas Municipal Court for men arrested for soliciting prostitution. Like others in the state who are close to the brutal physical and emotional damage of prostitution, Meadow told me that he is outraged at the lack of funds for services, pointing out that while Nevada ranks high in suicides, he said it is among the least funded in the United States for substance abuse and mental health services.[590] There is also a need for research on what many service providers and advocates reported to me: a frighteningly high suicide rate among young women in Nevada.

Nevada service providers include rape crisis centers, domestic violence programs and shelters, homeless shelters, clinics, and advocacy groups. There is consensus among these agencies that services for women who want to escape both legal and illegal prostitution as well as services for trafficked women escaping prostitution are urgently needed. The dilemma is that legal prostitution itself is a barrier to services for women escaping prostitution. When prostitution is considered a legal job instead of a human rights violation, why should the state offer services for escape? When prostitution is legal, there is a failure of state and local agencies to acknowledge and redress its harms, therefore funding is not allocated for necessary services. This same lack of services occurs in other countries where prostitution is legalized. For example, Sullivan notes that despite the Australian state of Victoria's legal responsibility to provide exit services for women in legal prostitution, no such programs were ever created.[591]

Once offered emotional support, medical care, and job alternatives that pay enough for a decent home, most women would immediately leave prostitution. Most women - 81% of those we interviewed - in the Nevada legal brothels want to escape prostitution. The longer a woman has been in prostitution, the more complex and costly her medical and psychological care becomes.

Alesia Adams joked about the lack of services for prostituted teens in Atlanta. "What do I have to do to get some help here? Take these girls across the border in a bus, then bring them back into the U.S., and say, 'Here, I've got some trafficked girls, now can we get some funding for services?'"[592] Adams was challenging the lack of funds for women and children who are domestically, but not internationally, trafficked into prostitution.

There are some excellent agencies in the United States that offer services for women and girls escaping prostitution. For example, Girls' Educational and Mentoring Services (GEMS) offers services to young women escaping prostitution.[593] GEMS emphasizes that girls arrested for prostitution are not criminals but instead are children who are sexually assaulted by predatory johns even if the john uses the perennial line "Gee, officer, she sure looked over 18 to me" and even if he says she 'consented.' GEMS and other groups are working with other New York to pass the Safe Harbor for Exploited Youth law which would offer children safe houses, medical care and support services rather than locking them up.[594]

While there are some services in the state for teenagers escaping prostitution, in 2006 there were no shelters in Nevada specifically for adult women escaping legal or illegal prostitution. There were no shelters specifically for trafficked women. Because of the high numbers of prostituted women and children in Las Vegas, law enforcement and service provision efforts have primarily targeted children to the exclusion of adult women.[595] These services are urgently needed. However, Nevada services for adult women escaping prostitution, who frequently need emergency housing, counseling, and medical treatment, are almost nonexistent. It's as if there's been an unspoken policy on the part of city fathers and law enforcement that if you refuse to offer services to women in dire straits – they'll get so frustrated and desperate that they'll somehow figure out a way to not only exit prostitution but also to leave the state of Nevada, thereby saving the state money.

Women escaping prostitution need secure shelter, medical and psychological care geared specifically to their needs, survivor-led support groups, economic assistance to access services, coordinated care, and vocational rehabilitation by those with expertise in sexual abuse and prostitution. Services for prostituted/trafficked women must be offered in many languages and in ways that are culturally relevant.[596] See Chapter 12 by

Jody Williams which describes the multiple barriers that women face when they attempt to escape prostitution in Nevada.

Advocates and policy makers may not adequately grasp the physical danger posed by pimps. A case manager described the coercive control of pimps over women and girls in Las Vegas, and the challenge of outreach:

> Girls don't have access to us, for example, we sent one caseworker out to speak to one girl in a Jack-in-the-Box [fast food restaurant], and her pimp wouldn't let her further than 20 feet away from him at all times, and then she disappeared.[597]

Clark County Sheriff Bill Young in 2003 expressed his frustration with Nevada's prostitution laws: "From a police perspective, it is almost impossible to stop from the lowest rung on up. Part of the reason is that soliciting for prostitution is just a misdemeanor crime, and so we may spend a lot of tax dollars to set up a sting and then the girl is out on bail two hours later."[598] Young's criticisms of the system make sense. If there is political will to address prostitution, then the law should reflect that it is a more serious crime than jaywalking.

In Australia, as in Nevada, the states and territories individually determine the legal status of prostitution. New research on Australian prostitution permits a comparison to that in Nevada. The Australian state of New South Wales (NSW) permits legal street prostitution.[599] Women in legal street prostitution in Australia had a 75% rate of childhood sexual abuse, and almost half of them had attempted suicide. In both locations, the economic savagery of governments that ignore the needs of the poor was clearly connected with prostitution. In NSW 45% of these legally prostituting women had been recently homeless. In the Nevada legal brothels, 47% of our interviewees had been homeless. Like the women in legal Nevada prostitution, the Australian women in legal street prostitution had high rates of lifetime trauma histories and current mental distress. 31% had been prostituted as children – a figure comparable to the 23% who reported prostituting as children in legal Nevada prostitution.[600] In the Australian study, the effects of racism increased the harms incurred by poverty and sexism. Indigenous Australian women significantly more often had been homeless (68%), and had significantly higher rates of depression when compared to the non-indigenous women.

As Louise Brown observed about prostitution in Sri Lanka, China, Vietnam, Burma, and the Philippines, "anyone passing through the red light areas of Asia's cities will be hard pressed to guess which legal system is in force."[601] The same contradictions are seen in Nevada, where legal prostitution is only 10% of all prostitution. All other prostitution in the state is illegal, but it is highly visibly advertised and in-your-face.

> What's really quirky is that while the legal brothels exist, it's absolutely hidden. It exists only in 10 mostly rural counties. It's not in any of the big urban centers. There is no legal advertising of brothels, there's very little to draw your attention to them. You have to stumble upon them. There's no legalized prostitution in Las Vegas. Yet, despite the fact that it's illegal, it is in your face everywhere here. *The truth is, where it's legal it's invisible and where it's illegal it's in your face.*[602]

Legal prostitution in other countries is also hidden. But the reasons for this are not simply quirkiness. The system of invisibility is carefully planned and implemented. As Claude Jaget explained about Morocco and France,

> They don't want to see us in their field of vision, in everyday life....We're like peoples' bad consciences...We have to exist, but at the same time, we have to be invisible. They want to imagine us hidden and dirty so they can despise us all the more, and then when they don't need us any more they can calmly wipe us from their memory.[603]

There are basically four legal approaches to prostitution. Prostitution can be criminalized, legalized, decriminalized, or abolished.[604] Where there is *criminalized prostitution,* all parties to the prostitution transaction are arrested: the woman who is selling sex in prostitution, the buyer or john, the pimp and the trafficker. *Legalized prostitution,* as in Nevada, specifies where prostitution is permitted to take place, and it is regulated via zoning. Legalized prostitution can also include municipal *tolerance zones or red-light zones*. *Decriminalized prostitution* removes all laws against pimping, pandering, and buying women in prostitution, in addition to decriminalizing the woman in prostitution herself. *Abolition of prostitution* is a human rights-based legal approach that aims to stop the buying, selling and trafficking of women in prostitution while *at the same time* supporting those in it to escape prostitution instead of arresting them. The abolitionist perspective considers the sexual harassment, sexual exploitation, and sexual violence of prostitution to be human rights violations. Victims in this fourth legal approach are the only ones who are decriminalized.

The fourth alternative to legalized prostitution is exemplified in the 1999 Swedish law that prohibits the purchase of sexual services.[605] Recognizing that prostitution deserved abolition and that it is a form of male violence against women, the Swedish Parliament decided to criminalize those who create and profit from systems of prostitution: not only the pimps and traffickers, but also those who purchase and exploit women and others who are used in prostitution: the buyers or johns. The woman or man in prostitution is not criminalized since Sweden has recognized that she is a victim rather than a criminal.[606] The Swedish law states that *"in the majority of cases… [the woman in prostitution] is a weaker partner who is exploited,"* and it allocated funding for social services to *"motivate prostitutes to seek help to leave their way of life."*[607]

Unless a country has an abolitionist antitrafficking law, like Sweden's,[608] it does not really make a difference to the organized criminals and small time pimps whether or not a country has legal or illegal prostitution. Pimps know that there will be little if any legal interference to selling women indoors, inside legal brothels, inside illegal brothels, out of escort agencies, massage parlors, at adult industry shows, and at nail parlors.

Gunilla Ekberg, who played an important role in the implementation of the 1999 Swedish law, told me that the law functions as a barrier to the marketing and third-party profiting from prostitution. "After eight and half years," Ekberg said, "the law is working." Although young women in other countries are being recruited into prostitution in record numbers, the number of newly recruited women in Swedish prostitution is low, and the overall number of women used in prostitution in Sweden has remained stable, rather than increasing as has occurred in many European countries where the sex industry is rapidly expanding.[609] The Swedish National Board of Health and Welfare, reported a 50% decrease in the number of women prostituting and a 75% decrease in the number men who bought sex.[610] Sex trafficking into Sweden has dramatically decreased since the law went into effect. In 2007 Sweden has the lowest number of victims of trafficking in the European Union. The law has also effectively reduced organized crime networks in Sweden.

The rationale underpinning the Swedish law is that prostitution is a harmful social institution deserving of abolition because prostitution always exploits those who are the most vulnerable and the most marginalized. We see this same analysis of prostitution in the statements of advocates on behalf of women and children in other countries. In India, opposition to legal prostitution was based on the argument that it essentially gives families[611] and men the license to legally exploit and abuse females. Lalitha Nayak of India's Joint Women's Programme argued that removing restrictions on prostitution will not halt abuses, instead it only encourages abuses.[612] Similarly, Australian professor Helen Pringle said of her opposition to legal prostitution in New South Wales: "Prostitution is not a profession but a subjugating verb."[613]

Parliamentarians, feminist groups, and non-governmental organizations in other European countries are considering laws that would criminalize men who buy women in prostitution, as Sweden did. Given what we know about the overwhelming harms of prostitution, it is crucial that laws are not compromised by addressing only the harms to trafficked women but not harm perpetrated against locally prostituted women. Unfortunately Finland recently passed such a flawed law.[614] On the other hand, stating that "Norway shall not be a free zone for pimps and human traffickers," Justice Minister Knut Storberget proposed a new law that would prohibit the purchase of sex, punishable with up to six months in jail. Like the Swedish law, the proposed 2007 Norwegian law will target customers, not their victims.[615]

In 2007, the New York legislature moved forward in the passage of a strong antitrafficking law. The law was crafted after that of Sweden, and several of the key features of the Swedish law are now part of the New York law, including increasing penalties against men who buy women in prostitution, increasing penalties for traffickers, and provision of services for victims of these predators.[616]

Two international agreements confront the flawed notion that women "freely consent" to prostitution, and make clear statements opposing prostitution and trafficking. The first is the United Nations 1949 Convention which declares that trafficking and prostitution are incompatible with individual dignity and worth.[617] The 1949 UN Convention addresses the harms of prostitution to *consenting adult women whether or not they were transported across national boundaries.* A second United Nations document views trafficked women as victims, not criminals. The 2000 Palermo Protocol like the 1949 Convention declares that consent is irrelevant to whether or not trafficking has occurred. It encourages states to develop legislative responses to men's demand for prostitution[618] and establishes a method of international judicial cooperation that would permit prosecution of traffickers and organized criminals. The Palermo Protocol also addresses a range of other forms of sexual exploitation including pornography.

Some not-so-favorable developments have also occurred in Nevada. Although legalization of prostitution is a failed social experiment - decriminalizing prostitution is certainly not a preferable alternative; it is an equally flawed response to the exploitation and abuses that are intrinsic to prostitution/trafficking. However Las Vegas sociologists Brents and Hausbeck advocated decriminalization of prostitution - a radical removal of all laws regarding prostitution including removal of prohibitions against pimping, pandering, and buying women.[619] Decriminalization of prostitution may at times be rooted in a well-meaning desire to help women in prostitution, but when johns and pimps are decriminalized, the sex industry itself, although inseparable from exploitation, is mainstreamed and legitimized.

Some professional groups and organizations have chosen to avoid Nevada for their meetings because of the state's legalized prostitution. State Assemblywoman Sheila Leslie said she thinks Nevada has been "penalized for our legalized prostitution." Leslie, who objected to prostitution because it exploits women, said her attempts to get the National Network for Youth to hold its conference in Reno failed because of the group's objection to the state's lax prostitution laws.[620]

Nevada and Las Vegas public officials are well aware of the economic importance of the sex industry to their state. Conventions are the lifeblood of Las Vegas. The refusal of groups and organizations to hold meetings in Las Vegas until the state removes the law giving counties the right to legalize prostitution and until Nevada replaces that law with an antitrafficking law criminalizing johns who buy women in prostitution (whether they are transported from California or Colombia) but not criminalizing the women in prostitution themselves - could have a serious economic impact. It could generate strong motivation for positive social and legal change.[621]

But some positive developments have occurred in Nevada. In the past, the testimony of a prostitute against a pimp required corroboration. Las Vegas Police Sergeant Gil Shannon successfully lobbied for a change in the Nevada law so that pimps could be arrested and prosecuted solely on the testimony of a prostitute.[622]

Suggesting that the Nye County law legalizing prostitution did nothing to change the essential harmfulness of prostitution or the illegal and abusive practices of pimps, Nye County Commissioner Trummell stated, "The way out of this is for Nevada to decide to be like the rest of the United States..." [623] Since leaving office, Candice Trummell and others have organized the Nevada Coalition Against Sex Trafficking (N-CAST) whose mission is to educate Nevadans about the human rights violations in prostitution, about the links between prostitution (both legal and illegal) and sex trafficking, to support survivor networking and to promote services that women need in order to escape or avoid prostitution, and to develop and promote policy and functional alternatives to current laws and practices within the State of Nevada.[624]

Legal prostitution in Nevada is like legal prostitution elsewhere. When prostitution is legalized, pimps and traffickers are attracted to the legal market and to the entire cultural milieu where prostitution is welcomed. Both legal and illegal prostitution increase, as does prostitution (sexual abuse) of children. Trafficking increases under legal prostitution.[625] All of these adverse effects have been observed in Australia, where after prostitution was legalized in Victoria, the number of legal brothels doubled. But the sex markets' greatest expansion was in the *illegal* prostitution sector. In a single year after legalization, there was a 300% growth of illegal brothels.[626] In addition to increasing illegal prostitution, the legalization of prostitution in Australia and in Nevada is associated with increases in the prostitution of children and in sex trafficking.[627]

Nevada's *illegal* prostitution industry is 9 times greater than the state's legal brothels. This is comparable to what has happened in the Netherlands and in Victoria, Australia, where the vast majority of the state's sex industry "operates either on the margins of the legal sector or in a totally illegal environment."[628] In Australia as in Nevada, the industry's boundaries are constantly blurring and expanding. Legislators are always many steps behind pimps and traffickers in struggling to regulate an industry that is impossible to regulate. One politician for example, told me that he did not know what the current ordinance regarding lap dancing is in Las Vegas, since it changes approximately monthly. New variations on the business of sexual exploitation and its hydra-headed global trafficking networks are constantly emerging.

Like legal prostitution elsewhere in the world, Nevada's legal brothels have provided a welcoming legal and cultural backdrop for the transport of women and children from other US states and from other countries both to the legal brothels and to Las Vegas and Reno, where prostitution is illegal. Given the facts described here, legal prostitution in Nevada is tantamount to state-sponsored exploitation and abuse.

All prostitution causes harm to women. Whether you are being sold by your family to a brothel; or whether you are being sexually abused in your family, running away from home, and then being pimped by your boyfriend; or whether you're in college and need to pay for next semester's tuition and you work at a strip club behind glass where men never actually touch you; whether you live in town and prostitute or whether you were moved half way around the world by traffickers to prostitute; whether you're paid $5 or $5000 for a blowjob - all prostitution causes harm to the women in it.

Ending the legalization of prostitution in Nevada would be a first step toward ending the harms of prostitution and the domestic and international traffic in women in the state. Enforcing existing laws against buying people for sex (the john's solicitation of the prostitute), and increasing the penalty for buying sex from a misdemeanor to a felony would be a crucial second step toward ending the sexual exploitation of prostitution. Finally, it is essential that throughout the slow process of social and legal change, women and men and the transgendered who seek to escape prostitution and trafficking are provided immediate and longterm shelter, medical treatment, social services, and vocational rehabilitation with job placement and support.

Appendix A.

Legal Status of Prostitution by Each of Nevada's 17 Counties in 2007

	Prostitution is Illegal	Prostitution is Legal	Prostitution has no Legal Status
Carson City	Yes		
Douglas	Yes		
Lincoln	Yes		
Clark[1]	Yes		
Washoe[2]	Yes		
Elko		Yes	
Humboldt		Yes	
Pershing		Yes	
White Pine		Yes	
Churchill		Yes	
Esmeralda		Yes	
Lander		Yes	
Lyon		Yes	
Mineral		Yes	
Nye		Yes	
Eureka[3]			Yes

(1) Las Vegas is in Clark County
(2) Reno is in Washoe County
(3) Nevada state law leaves it up to the counties to determine the legal status of prostitution. Eureka County neither permits nor prohibits brothels.

Appendix B.

2007 Las Vegas Advertising Costs

The data in Table 6 summarizing Las Vegas-based costs for print, magazine rack, cards, flyers, telephone book, and online advertising was a research project in and of itself. Kathy Watkins was the primary investigator for the research on Las Vegas advertising costs, a project that took approximately 6 months to complete. Some agencies were extremely reluctant to provide us with costs of their services - hinting that until we were ready to run an ad, they wouldn't provide the information we sought.

The numbers in Table 6 are conservative. Actual advertising costs are likely to be much higher. We included data from only 4 sources. A more comprehensive survey of prostitution advertising costs - which might have tripled or quadrupled the costs reported here - would also include: international search engines, online message board advertising, Las Vegas-based print magazines, U.S.- and internationally-based magazines, and more than one Yellow Pages telephone directory.

Print advertising rate estimates were based on rates found at www.psprint.com. Of the 2667 registered magazine racks in Clark County, 2663 contain escort prostitution advertising. We estimated that each magazine rack distributed 200 print catalogs each per week for a total of total of 2,263,550/month. We estimated a cost of $0.331 per print catalog based on rates from psprint.com. The annual fee for renting a magazine rack is determined by Clark County.

Flyer advertising rate estimates were based on rates found at Based on advertised prices from www.konceptkreations.com. We estimated that at least 2 million 2.5 x 4 inch cards per month at $.0147 each were distributed on the Strip and other locations in Las Vegas. Cost estimates are based on rates quoted by Modern Postcard.

Our estimate of **Internet advertising** is conservative and is based on an estimate of 760 escort prostitution advertisements on 3 online escort prostitution directories at a rate of $20 per week for individuals providing escort prostitution and $60 per week for agencies providing escort prostitution. We included building, maintaining, and hosting costs for 360 small websites at $125 per month.

Yellow Pages advertising rates are based on a quote provided via telephone in July 2007 by Embarq Yellow Pages, located at 8311 W. Sunset Road, Suite 250, Las Vegas, 89113. See also www.bestredyp.com. We obtained rates and estimated costs for one of the telephone directories in Las Vegas, but we are aware of at least one additional Yellow Pages directory. The cost of Yellow Pages advertising by color and size are as follows:

In 2007, the U.S. District court struck down a ban on legal brothel advertising in Nevada counties that do not permit legal prostitution. As a result, it is likely that prostitution advertising will significantly increase. See Ray Hagar (2007) Counties without prostitution must permit advertisements. (*Reno Gazette-Journal* July 14, 2007. Retrieved July 20, 2007 from http://news.rgj.com/apps/pbcs.dll/article?AID=/20070714/BIZ14/70

Yellow Pages Ad size	# of ads	Rate per month	Total per month	Total per year
Full page Full color	51	6641	338,691	4,064,292
Full page – white & black	10	6037	60,370	724,440
Full page – black only	50	4678	233,900	2,806,800
3/4 page – full color	8	5439	43,512	522,144
3/4 page – white & black	2	5088	10,176	122,112
3/4 page – black only	7	3766	26,362	316,344
1/2 page – full color	4	4713	18,852	226,224
1/2 page – white & black	2	3924	7,848	94,176
1/2 page – black only	5	2854	14,270	171,240
1/4 page – full color	4	3063	12,252	147024
1/4 page – white & black	2	2551	5,102	61224
1/4 page – black only	6	1855	11,130	133560
1/8 page – full color	13	1279	16,627	199524
1/8 page – white & black	12	1151	13,812	165,744
1/8 page – black only	53	806	42,718	512,616
1/16 page – full color	3	959	2,877	34,524
1/16 page – white & black	38	863	32,794	393,528
1/16 page – black only	268	605	162,140	1,945,680

Endnotes

Foreward

[1] http://www.klas-tv.com/Global/story.asp?S=6421545.

[2] http://www.lasvegassun.com/sunbin/stories/lv-other/2007/jan/29/566684903.html.

[3] Washingtonian, November 2005, "X Rated," by C.J. Vogel. http://www.washingtonian.com/articles/mediapolitics/4017.html.

[4] Central Intelligence Agency Directorate of Intelligence (2007) Global Human Trafficking: Updated Estimates (Unclassified) March 8, 2007.

[5] Reno Gazette-Journal (2007) "Blaze marks end of Mustang Ranch II" http://www.bunnyranch.com/news/RGJ_com%20Blaze%20marks%20end%20of%20Mustang%20Ranch%20II.htm. March 26, 2007.

[6] Brian Haynes (2007) "Task force started." Las Vegas Review-Journal February 7, 2007. http://www.reviewjournal.com/lvrj_home/2007/Feb-07-Wed-2007/news/12437732.html.

[7] http://assembly.state.ny.us/leg/?bn=A08679&sh=t.

[8] http://www.leg.state.nv.us/74th/Bills/AB/AB383_R2.PDF#xml=http://search.leg.state.nv.us/isysquery/irle990/17/hilite

[9] Mark Lagon (2007) Speech from Office to Monitor and Combat Trafficking in Persons, U.S. State Department.

Chapter 1. Introduction

[10] United Nations (2006) Office of Drugs and Crime Report on "Trafficking in Persons: Global Patterns" (2006), page 33. 87% are trafficked for sexual exploitation and 28% are trafficked for labor exploitation. The total percentage exceeds 100% because people experience more than one kind of exploitation. Retrieved January 4, 2007 from http://www.unodc.org/unodc/en/trafficking_persons_report_2006-04.html.

[11] Joel Brinkley (2006) The Saturday Profile: A Modern-Day Abolitionist Battles Slavery Worldwide. New York Times. May 24, 2007. Retrieved June 2, 2007 from http://select.nytimes.com/search/restricted/article?res=F00A15F83D5A0C778CDDAB0894DE404482. See also Nicholas D. Kristof (2006) Bush Takes on the Brothels. New York Times May 9, 2006. Retrieved May 12, 2006 from http://select.nytimes.com/search/restricted/article?res=F30B1FFD3D5A0C7A8CDDAC0894DE404482.

12 Prostitution Research & Education (PRE) located in San Francisco, was established in 1995. See www.prostitutionresearch.com. PRE is a 501(c) 3 nonprofit organization that conducts research on prostitution, pornography and trafficking, and offers education and consultation to researchers, survivors, the public and policymakers. PRE's goal is to abolish the institution of prostitution while at the same time advocating for alternatives to trafficking and prostitution - including emotional and physical healthcare for women in prostitution. The root of the problem of trafficking for prostitution is men's demand for prostitution. Emphasizing the roots of prostitution and trafficking in racism and poverty as well as lethal sexism, PRE collaborates with other organizations in all projects.

13 J. Banks (2003) Comment: City Shoulders Load of Making Law Work. Auckland: *New Zealand Herald* September 15, 2003.

14 Lorraine Bonner (2007) The Perpetrator Series. *Eleventh Annual San Francisco Artists Against Rape.* San Francisco Women Against Rape. In discussing her work, sculptor Lorraine Bonner explained the connection between disbelief and denial which formed the basis for my ideas in this paragraph.

15 Brooke Adams (2007) Polygamist leader Jeffs' appeals rejected by Utah Supreme Court: Justices deny petition to move venue, void the states's rape statute. *Salt Lake Tribune* July 6, 2007. Retrieved July 6 2007 from http://www.sltrib.com/news/ci_6310152.

16 Mary Lucille Sullivan (2007) *Making Sex Work: A Failed Experiment with Legalized Prostitution,* page 4. North Melbourne:Spinifex.

17 Dissociation is a mental tuning-out that protects the psyche from overwhelming traumatic stress. Dissociation occurs among prisoners of war who are tortured, among children who are sexually assaulted, and among women who are battered, raped, or prostituted. Dissociation is common among women in street, escort, and strip club prostitution. It is a result of both childhood sexual violence and sexual violence in adult prostitution. At the same time, dissociation is a job requirement for surviving prostitution.

Most women report that they cannot prostitute unless they dissociate. When they do not dissociate, they are at risk for being overwhelmed with pain, shame, and rage. One woman explained, "It's almost like I trained my mind to act like I like prostitution but not have any thoughts. I have the thoughts like 'What is this doing to my body and my mind and my self-esteem?' a few days later but not as it's happening....Even though the guys are paying me for it, I feel like they're robbing me of something personal. And I wonder, 'Why are they doing this?'" Virginia Vitzthum, Selling Intimacy, *Salon.com* July 25, 2000. Retrieved September 22, 2004 from http://archive.salon.com/sex/feature/2000/07/25/girl_part_iii/index1.html (quoting a prostituted woman). See also Marianne Wood, *Just a Prostitute* (1995) discussing the rage that is a consequence of tolerating johns' behaviors.

[18] Canada, Colombia, Germany, Mexico, New Zealand, South Africa, Thailand, Turkey, United States, and Zambia. In Germany, prostitution in brothels is legal and in New Zealand, brothel prostitution is decriminalized. Prostitution is illegal in the other countries, but enforcement of laws against it is often inconsistent. See Melissa Farley, Ann Cotton, Jacquelyn Lynne, Sybile Zumbeck, Frida Spiwak, Maria E. Reyes, Dinorah Alvarez , Ufuk Sezgin (2003) Prostitution and Trafficking in 9 Countries: Update on Violence and Posttraumatic Stress Disorder. *Journal of Trauma Practice* 2 (3/4): 33-74. Also available at http://www.prostitutionresearch.com/c-prostitution-research.html.

[19] Barbara G. Brents and Kathryn Hausbeck (2005) Violence and Legalized Brothel Prostitution in Nevada: Examining Safety, Risk and Prostitution Policy. *Journal of Interpersonal Violence* 20 (3), page 275.

[20] John Lowman (2005) Speaking at 38th session of the Canadian Parliament, Subcommittee on Solicitation Laws of the Standing Committee on Justice, Human Rights, Public Safety and Emergency Preparedness. Feb 21, 2005. Retrieved March 7, 2005 from http://www.parl.gc.ca/committee/CommitteePublication.aspx?COM=0&SourceId=103606&SwitchLanguage=1.

[21] Gabriel R. Vogliotti (1975) *The Girls of Nevada*. Secaucus, NJ: Citadel Press, page 251. A woman in prostitution explained, "Of course, the laws regarding prostitution are not primarily intended to protect me. The fact that we can only work in state-licensed houses keeps the brothel owners in business and protects the neighbors from whores running loose in their streets and backyards. State regulations benefit everyone but the working girl." Retrieved Jun3, 2006 from http://www.headlightjournal.com/essays/whore/whore.html.

[22] Kathleen Barry (1995) *The Prostitution of Sexuality*. New York University Press, page 228.

[23] Micloe Bingham (1998) Nevada Sex Trade: a gamble for the workers. *Yale Journal of Law & Feminism*, pages 69-99.

[24] Sullivan, *Making Sex Work*, pages 57-58.

[25] Vogliotti, *The Girls of Nevada*, page 7.

[26] When coltran, a metal used in used in the superconductor chips that make cell phones work, was located in the Ituri peoples' land in eastern Congo, the miners worked under brutal conditions in muddy water. They were provided with prostitutes for pacification at the end of the long days. Because of environmental protests the mines were shut down, but the prostitution continued. Blaine Harden (2001) A Black Mud From Africa Helps Power the New Economy. *New York Times Magazine* August 12, 2001.

[27] Jeanie Kasindorf (1985) *The Nye County Brothel Wars*. New York: Dell, page 28.

[28] Las Vegas is located in Clark County.

[29] Kasindorf, *Nye County Brothel Wars*, pages 28-9. The population limit has been revised upward to 400,000.

[30] One study found that children were more likely to be sexually abused in prostitution when there was either a military base or a thriving sex industry close by. Richard J. Estes and Neil A. Weiner (2001) Commercial Sexual Exploitation of Children in the U.S.. Canada, and Mexico. University of Pennsylvania. Nellis Air Force base, located eight miles from Las Vegas, is the size of Connecticut. Nellis has a population of 12,000 people.

[31] Nevada is experiencing a boom in gold and copper mining in the northern part of the state.

[32] Nevada Natural Resources Status Report. Retrieved July 9, 2006 from http://dcnr.nv.gov/nrp01/land01.htm.

[33] David Loomis (1993) *Combat Zoning: Military Land-use Planning in Nevada*. Reno: University of Nevada Press, page vi. See also photos of 80,000 Navy-controlled acres in northern Nevada that are beneath 6 million acres of military airspace at http://www.clui.org/clui_4_1/lotl/v27/k.html.

[34] Loomis, *Combat Zoning*, pages 5, 11.

[35] Nevada page of Roger J. Wendell. Retrieved May 5, 2007 from http://www.rogerwendell.com/nevada.html.

[36] Peggy Reeves Sanday (1981) The Socio-Cultural Context of Rape: A Cross-Cultural Study. *Journal of Social Issues* 37 (4): 5-27.

[37] See the legal status of prostitution in Nevada's 17 counties in Appendix A. State archivist Guy Rocha described a period in Nevada's history (1970) when the Bureau of Land Management rented land in Esmeralda County to a brothel owner, in effect making the Secretary of the Interior a pimp. Federal action was quickly taken and the pimp was evicted. Guy Louis Rocha (1999) *Nevada's Most Peculiar Industry: brothel prostitution, its land use implications and its relationship to the community*. Nevada State Library and Archives. Paper on file with author.

[38] Personal Communication (2006) with Candice Trummell. Phone interview July 21, 2006.

[39] FBI (2006) Pahrump brothel owner arrested and charged with fraud for unlawful payments to Nye County Commissioner. Retrieved March 8, 2006 from http://lasvegas.fbi.gov/dojpressrel/pressrel06/wirefraud030606.htm. See also Henry Bean (2006). Ex-Official sees Pattern of Bribery: Former Commissioner says he told FBI about Brothel Owner in '90's. *Las Vegas Review-Journal* March 8, 2006. Retrieved March 10, 2006 from http://www.reviewjournal.com/lvrj_home/2006/Mar-08-Wed-2006/news/6248782.html.

[40] Kasindorf, *Nye County Brothel Wars*, page 33. Kasindorf's courageous and invaluable book about corruption and violence associated with legal brothel prostitution, *The Nye County Brothel Wars*, details the history of one county's legal and physical battles over prostitution, including pimp turf wars, and political corruption investigations resulting in convictions against two Nevada state senators, a Reno city councilman, the director of the Clark County business license bureau, two Clark County Commissioners and the first federal judge in

the history of the United States ever convicted of a felony committed while sitting on the bench (Kasindorf, *Nye County Brothel Wars*, page 301). A detailed discussion of the millennial version of similar political corruption follows in a later section of this book.

[41] There are a few more or a few less than 30 brothels at various times. Brothels regularly go into bankruptcy; others have been shut down by the health department. For a list of the legal status of prostitution in Nevada's 17 counties see Appendix A or www.prostitutionresearch.com/Nevada for updates.

[42] *Cathouse* is based on a brothel in northern Nevada visited by the author. HBO's *Cathouse* was described by critic Patrick Bromley as "cheap, dishonest art." The series fails to address the reality of life in Dennis Hof's brothel. It avoids any analysis of what is really happening between pimp, assistant pimp/madam, prostitutes and johns. There is "play-acting on both sides of the camera. For a channel that prides itself on... honest programming, HBO ...wasted an opportunity with Cathouse—their coverage amounts to little more than an advertisement for [a brothel]." Retrieved June 24, 2006 from http://www.dvdverdict.com/reviews/cathouse.php.

[43] See Lily Burana (2001) *Strip City: a stripper's farewell journey across America*. New York: Hyperion, page 206.

[44] Lenore Kuo (2002) *Prostitution Policy: Revolutionizing Practice through a Gendered Perspective*. New York: New York University Press, page 80.

[45] Douglas Lindeman (2000) *Angel's Ladies*. Lindeman produced and directed this documentary film. Berkeley: Film Kitchen. For information about the documentary, contact distributor Orly Ravid, at orlyravid@comcast.net. See also Alexa Albert (2001) *Brothel: Mustang Ranch and Its Women*. New York: Random House, pages 198-199. Albert's book is a personal and informative account of time spent in a Nevada brothel. Albert conducted extensive interviews with some of the women in the brothel. As we did, Albert found the women to be honest and generous with respect to the information they provided.

Chapter 2. Legal Brothel Prostitution in Nevada

[46] Adam Tanner (2006) Nevada gives legalized prostitution uneasy embrace. February 13, 2006. Reuters UK. Retrieved March 13, 2006 from http://today.reuters.co.uk/news/newsArticle.aspx?type=reutersEdge&storyID =2006-02-13T144614Z_01_ZWE353157_RTRUKOC_0_LIFE-PROSTITUTION.xml.

[47] There are regular skirmishes between mainstream business interests and brothel owners regarding the locations of brothels which some fear will scare off legitimate business in the state. See Susie Vasquez (2002) Dermody files suit to halt brothel construction in northern Storey County. *Nevada Appeal* February

2, 2002. Accessed March 10, 2006 from http://www.nevadaappeal.com/apps/pbcs.dll/article?Date=20020202&Catego ry=NEWS&ArtNo=202020103&Ref=AR&template=printart.

[48] Kathryn Hausbeck and Barbara G. Brents (2000) Inside Nevada's Brothel Industry, page 230. In Ronald Weitzer (ed.) *Sex for Sale: Prostitution, Pornography and the Sex Industry*. New York: Routledge, pages 217-241.

[49] Terri Miller (2005) Meeting at Nevada Coalition Against Sexual Assault office. June 29, 2005, Henderson, Nevada. Miller was Member and Training Services Coordinator of NVCASA at the time.

[50] Prostitution is not however culturally promoted as a job option for poor men.

[51] For this report we interviewed 45 women in legal brothels and found that 81% of them wanted to escape prostitution. In a comparable study of prostitution/trafficking in 9 countries, 89% of 854 people in prostitution wanted to escape prostitution (Melissa Farley, Ann Cotton, Jacquelyn Lynne, Sybile Zumbeck, Frida Spiwak, Maria E. Reyes, Dinorah Alvarez, Ufuk Sezgin (2003) Prostitution and Trafficking in 9 Countries: Update on Violence and Posttraumatic Stress Disorder. *Journal of Trauma Practice* 2 (3/4): 33-74. Also available at http://www.prostitutionresearch.com/c-prostitution-research.html. Also in Melissa Farley (2003) (ed.) *Prostitution, Trafficking and Traumatic Stress*. Binghamton: Haworth, pages 33-74.

[52] Robert Longley (2004) Gender Wage Gap Widening, Census Data Shows. *U.S. Government Information/Resources*. September 1, 2004. Retrieved July 7, 2006 from http://usgovinfo.about.com/od/censusandstatistics/a/paygapgrows.htm.

[53] Ian Urbina (2006) Keeping It Secret As the Family Car Becomes a Home. *New York Times* April 2, 2006. Study conducted by the National Low Income Housing Coalition. Retrieved June 17, 2006 from http://www.nytimes.com/2006/04/02/us/02cars.html?ex=1144641600&en=3 36986e591dbeef2&ei=5070&emc=eta1.

[54] Helen Reynolds (1986) *The Economics of Prostitution*. Springfield: Charles C. Thomas, page 13. Because of the wage differential between men and women, women are more adversely affected by inflation than men.

[55] Mary Lucille Sullivan (2007) *Making Sex Work: A Failed Experiment with Legalized Prostitution*. North Melbourne: Spinifex, page 163. Quoting Senate Community Affairs Inquiry into Poverty in Australia, 2003.

[56] Lois Rita Helmbold (2005) Women's Studies in Sin City: Reactionary Politics and Feminist Possibilities. *National Women's Studies Association Journal* 17 (2): 171-177. Summer 2005. Quoting Amy B. Caiazza and April Shaw (eds.) (2004) *The Status of Women in Nevada*. Washington D.C.: Institute for Women's Policy Research, pages 52–67.

[57] Nevada's 2004 suicide death rate was 19.2 per 100,000 population. California's rate was 9.6 and New York's 6.0. In 2003, the U.S. suicide death

rate was 10.8 Retrieved January 8, 2007 from http://www.cdc.gov/nchs/Default.htm.

[58] In 2004 Nevada's suicide death rate for youth was 6.8 deaths per 100,000 youth. The U.S. average was 4.7 in that same time period. Data courtesy of Jodi Tyson, Youth Suicide Prevention Program Coordinator, Nevada Office of Suicide Prevention, Department of Health and Human Services, Las Vegas. January 16, 2007.

[59] Douglas Lindeman (2000) *Angel's Ladies*. Lindeman produced and directed the documentary film. Berkeley: Film Kitchen. For information, contact distributor Orly Ravid, at orlyravid@comcast.net.

[60] Richard R. Becker and Ellen Levine (1994) Taking the Sin out of Sin City: a look at prostitution in Nevada. *Gauntlet* 1: 33-40, page 36.

[61] National Law Center on Homelessness and Poverty and National Coalition on Homelessness. (2006) *A Dream Denied: the Criminalization of Homelessness in U.S. Cities*. The criteria used for ranking cities include: number of anti-homeless laws in the city, enforcement and severity of penalties, political climate toward the homeless, local homeless advocates' support for or opposition to the designation of being a "mean city," history of criminalization measures, and existence of pending or recently enacted criminalization legislation in the city.

[62] Kathleen Hennessey (2006) Feeding Homeless is a Losing Bet in Las Vegas. *Deseret News* December 10, 2006. Retrieved January 19, 2007 at http://deseretnews.com/dn/view/0,1249,650213915,00.html. See also Kathleen Hennessey (2006) Glitzy Las Vegas battles over homeless. Associated Press. December 18, 2006. Retrieved January 9, 2007 from http://news.yahoo.com/s/ap/20061218/ap_on_re_us/homeless_in_vegas.

[63] See advertisement from May 2005 *Vanity Fair*, page 145. Captioned "What happens here, stays here," the ad pictures a duct tape roller like those commonly used to remove cat or dog hair from clothes and furniture. The roller has stuck on it the trappings of prostitution - her feathers, sequins, eyelashes and even an earring are stuck to the roller. The message is that with a few whisks, she can be disappeared, removed from the picture.

[64] Erin Neff (2003) Legalized Prostitution: Vegas brothels suggested - Goodman remarks open debate on downtown red-light district. *Las Vegas Review-Journal* October 24, 2003. Retrieved December 2, 2005 from http://www.reviewjournal.com/lvrj_home/2003/Oct-24-Fri-2003/news/22438503.html.

[65] Vancouver Canada Police Department. 2004 Interview with Detective Constable Raymond Payette, Youth Squad-KEYS. Terri Miller at the Las Vegas Metro Police Department similarly noted that international demand and legalized prostitution set up a "ripe environment for human trafficking." Lynette Curtis (2007) Activist again in league of her own, police unit battles trafficking. *Las Vegas Review-Journal* May 21, 2007. http://www.lvrj.com/news/7606732.html.

[66] Dennis Hof (2007) speaking on KGO radio, San Francisco. July 2007.

[67] Jebbie Whiteside (2004) The Invisible Woman: Domestic Violence and Nevada Prostitution. Unpublished paper on file wih author, page 12. See also Christine Stark and Carol Hodgson (2003) Sister Oppressions: Prostitution and Wife Battering. In Melissa Farley (ed.) *Prostitution, Trafficking and Traumatic Stress* (2003) Binghamton: Haworth, pages 17-32.

[68] Decriminalized prostitution is a radical legal approach that mainstreams prostitution to the point where no laws exist pertaining to prostitution. Under decriminalized prostitution, the buying and selling of women in prostitution is the legal equivalent of buying and selling cigarettes.

[69] A. Else (2003) Opinion. *New Zealand Forward Sunday Supplement* July 6, 2003.

[70] Once officially registered as prostitutes, Dutch women feared that this designation would pursue them for the rest of their lives. Despite the fact that if they officially registered as prostitutes, they would then accrue pension funds, the women still preferred anonymity.

[71] Suzanne Daley (2001) New Rights for Dutch Prostitutes, but No Gain. New York Times August 12, 2001. Retrieved August 25, 2001 from http://www.nytimes.com/2001/08/12/international/12DUTC.html.

[72] Christine Stark (2004) "Girls to boyz: Sex radical women promoting pornography and prostitution. In Christine Stark and Rebecca Whisnant (eds.) *Not for Sale: Feminists Resisting Pornography and Prostitution.* North Melbourne: Spinifex, pages 278-291. Survivor account of M. Hanson, footnote 8, page 285.

[73] Kathleen Barry (1995) *The Prostitution of Sexuality.* New York: New York University Press, page 232

[74] Leah Platt (2001) Stopping at a Red Light. *The Nation* July 9, 2001. Retrieved June 7, 2004 from http://www.thenation.com/docprint.mhtml?i=20010709&s=platt. A survivor/writer described frequent "intimidation tactics" in the legal brothels. See also Ann d'Lorenzo (2001) A San Francisco Whore in a Nevada Brothel: Classiness and Class-Consciousness Clash in a Tale of Two Cities. Retrieved August 8, 2006 from http://www.headlightjournal.com/essays/whore/whore.html.

[75] Even people who promote prostitution as a reasonable job have described Nevada's legal brothels as prison-like. See for example Veronica Monet quoted in Christine Stark. (2004) "Girls to boyz: Sex radical women promoting pornography and prostitution," page 285.

[76] Claude Jaget (1980) *Prostitutes - Our Life.* Bristol: Falling Wall Press, pages 66-67.

[77] "What we have here is a captive audience, a *very* captive audience," said a door-to-door salesman who sold clothes to the Nevada brothels. Alexa Albert (2001) *Brothel: Mustang Ranch and Its Women.* New York: Random House, page 209.

[78] One teenager who escaped told us that she had not eaten for 5 days.

[79] See for example Richard Abowitz (2005) The Life. *Las Vegas Weekly* June 30, 2005. Retrieved September 8, 2005 from http://www.lasvegasweekly.com/2005/06/23/feat1.html.

[80] Lindeman, *Angel's Ladies*. Many brothels permit only a very few personal items in the rooms in which women both live and turn tricks. The brothel rooms have pornography tacked to their bedroom walls, and large plastic containers of lubricant, handiwipes, and condoms. Women who prostitute also keep lots of toothpaste and mouthwash on hand.

[81] Richard Abowitz (2001) Cathouse Dreams. *Las Vegas Weekly* June 7, 2001. Retrieved March 8, 2005 from http://www.lasvegasweekly.com/2001/features/06_07_CathouseDreams/cathouse_dreams.htm.

[82] Lenore Kuo (2002) *Prostitution Policy: Revolutionizing Practice through a Gendered Perspective.* New York: New York University Press, pages 81-82. Kuo cites information from Ellen Pillard (1983) "Legal Prostitution: Is It Just?" *Nevada Public Affairs Review* 2: 43-47.

[83] Prostitution Research & Education has begun a cross cultural study of men who buy women in prostitution. One man explained that prostitution was "like renting an organ for 10 minutes." Melissa Farley (2007) 'Renting an Organ for 10 Minutes:' What Tricks Tell us about Prostitution, Pornography, and Trafficking. In D. Guinn (ed.) *Pornography: Driving the Demand for International Sex Trafficking.* Los Angeles: Captive Daughters Media.

[84] Jaget, *Prostitutes – Our Life*, page 75.

[85] Laura Anderson (1995) Working in Nevada. Retrieved December 19, 2004 from http://lasvegas.about.com/gi/dynamic/offsite.htm?site=http%3A%2F%2Fwww.bayswan.org%2FLaura.html.

[86] The pimps do not always report their 50% cut of the women's prostitution earnings to IRS. We are aware of one case where 1099's for 100% of her earnings in legal prostitution were illegally ascribed to the woman herself, with the pimp failing to report his 50% of her earnings.

[87] Kuo, *Prostitution Policy*, page 83.

[88] One woman told us that the pimp who ran a Nevada brothel charged his women $10. for a cup of dried noodles that he purchased for $1. This man also had a restaurant in his brothel, and charged johns $5. for hamburgers. The women in the brothel were charged $10. for the hamburger. This practice has also been documented by Alexa Albert in *Brothel*, page 66.

[89] Louise Brown (2000) *Sex Slaves: The Trafficking of Women in Asia.* London: Virago, page 214.

[90] Jebbie Whiteside, The Invisible Woman: Domestic Violence and Prostitution, page 19.

[91] Interview with anonymous brothel employee, 2005.

[92] For a description of pervasive violence in different kinds of prostitution, including legal prostitution, see Melissa Farley (2004) "Bad for the Body, Bad for the Heart:" Prostitution Harms Women Even If Legalized or Decriminalized. *Violence Against Women* 10: 1087-1125. Prostitution can be and often is lethal. Potterat and colleagues found an extremely high homicide rate among prostituted women and observed that prostituted women "face the most dangerous occupational environment in the United States." John J. Potterat, Devon D. Brewer, Stephen Q. Muth, Richard B. Rothenberg, Donald E. Woodhouse, John B. Muth, Heeather K. Stites, and Stuart Brody (2004) Mortality in a Long-Term Open Cohort of Prostitute Women. *American Journal of Epidemiology* 159: 778-785.

[93] Roberto J. Valera, Robin G. Sawyer, Glenn R. Schiraldi (2001) Perceived Health Needs of Inner-City Street Prostitutes: A preliminary study. *American Journal of Health Behavior* 25: 50-59.

[94] Farley et al., Prostitution in 9 Countries, page 63.

[95] Barbara G. Brents and Kathryn Hausbeck (2005) Violence and Legalized Brothel Prostitution in Nevada: Examining Safety, Risk and Prostitution Policy. *Journal of Interpersonal Violence* 20 (3): 270-295, page 289.

[96] Brents and Hausbeck, Safety, Risk and Prostitution Policy, page 283.

[97] Daley, New Rights for Dutch Prostitutes, but No Gain.

[98] Brents and Hausbeck, Safety, Risk and Prostitution Policy, page 280.

[99] Associated Press (2004, April 16) Prostitute Sues Former Motley Crue Singer, Brothel Over Alleged Assault. Retrieved April 17, 2004 from http://www.krnv.com/Global/story.asp?S=1791978&nav=8faOMMmQ.

[100] Statement made in 1999 by a woman in a Nevada brothel resident to Indrani Sinha, Director of Sanlaap, an organization working with prostituted women and girls in India and Bangladesh. Interview with Indrani Sinha, Kolkata, May 2004.

[101] Personal Communication (2006) with Barbara Strachan, DIGNITY Services/Diversion Programs. Phoenix, Arizona. April 20, 2006

[102] Harvey L. Schwartz (2000) *Dialogues with Forgotten Voices: Relational Perspectives on Child Abuse Trauma and the Treatment of Severe Dissociative Disorders* (2000) New York: Basic Books, page 122.

[103] Tooru Nemoto, Don Operario, Mie Takenaka, Mariko Iwamoto, and Mai Nhung Le (2003) HIV Risk Among Asian Women working at Massage Parlors in San Francisco. *AIDS Education and Prevention* 15(3): 245-256.

[104] Jebbie Whiteside made valuable contributions to this research. She was instrumental in accessing interviewees both in and outside the brothels, and in conducting the brothel interviews. Whiteside was a law student at University of Nevada at Las Vegas during this research project. She wrote a paper while at UNLV describing the essential similarity between domestic violence and Nevada brothel prostitution: Whiteside, Jebbie. (2004) *The Invisible Woman: Domestic Violence and Prostitution*. Unpublished paper on file with author.

[105] Most brothels have no locks on the doors of the women's bedrooms/trick rooms. The lack of privacy in the brothels is intimidating and at the same time infantilizing.

[106] Peter Landesman (2004) The Girls Next Door. *New York Times Magazine.* January 25, 2004. Retrieved February 18, 2004 from http://www.nytimes.com/2004/01/25/magazine/25SEXTRAFFIC.html.

[107] Alexa Albert (2001) *Brothel: Mustang Ranch and Its Women.* New York: Random House.

[108] Jeanie Kasindorf (1985) *The Nye County Brothel Wars,* New York: Dell, offers similar cases that are typical of the lives of women prior to entry into legal prostitution. For example, one woman who prostituted in a southern Nevada brothel had been incested by her father. Her mother turned the child over to her father when she was eight, first applying Vaseline to the girl. Another woman was raped by her stepfather and brothers when she was thirteen, then turned out by them and emotionally coerced into prostitution.

[109] Sarah Katharine Lewis (2006) *Indecent: How I Make it and Fake it as a Girl for Hire.* Emeryville: Seal Press, page 239.

[110] Lewis, *Indecent,* page 287.

[111] Jaget, *Prostitutes- Our Life,* page 64.

[112] For a summary of this research, see Farley, 'Bad for the Body, Bad for the Heart,' pages 1087-1125

[113] Farley et al., Prostitution and Trafficking in 9 Countries, page 51.

[114] Other types of prostitution described by the women we interviewed included prostitution at truck stops, apartments, travel shows, boat shows, and bus stops.

[115] Melissa Farley, Jacquelyn Lynne, and Ann Cotton (2005) Prostitution in Vancouver: Violence and the Colonization of First Nations Women. *Transcultural Psychiatry* 42: 242-271.

[116] Inadequate emergency room data on possible or probable injuries from domestic violence including prostitution is an obstacle not only in Nevada– it is also a barrier to awareness of violence against women and service provision in many other states.

[117] 70% of Nevada's deaths by suicide occur in Las Vegas. Jodi Tyson MPH provided this and other information that was extremely important to this research. Tyson is currently working as Youth Suicide Prevention Program Coordinator at the Nevada Office of Suicide Prevention, Las Vegas.

[118] In an international study of prostitution, 75% of women reported having been previously homeless. Farley et al., Prostitution in 9 Countries, page 43.

[119] Numerous sources report that most women are pimped into the legal brothels including law enforcement victim advocates, women who currently prostitute in the Nevada brothels, women who previously prostituted in Nevada brothels, and advocates/service providers. A woman interviewed by Albert said, "God forbid if you ever called him a pimp. It wasn't even in your

vocabulary. It was like a bad word…But as far as I'm concerned, if you're sending your money to a man who wouldn't be with you if you weren't sending him money, then he's not your boyfriend, he's your pimp. Still, it took me a year after I left Bobby to be able to call him my pimp." (Albert, *Brothel*, page 76). Some pimps who placed women in the Nevada brothels staggered three-week brothel schedules among a stable of women so that only one would be home at any given time and the pimp could fool each woman into believing she was his only "girlfriend." (Albert, *Brothel*, page 78).

[120] Personal Communication (2005) with Paddy Lazar, former staff member of Council for Prostitution Alternatives, Portland, Oregon. July 1, 2005. See also Albert, *Brothel*, page 75.

[121] Several women showed us pimps' tattoos. See also Albert, *Brothel*, page 81.

[122] Whiteside, *The Invisible Woman: Domestic Violence and Prostitution*, page 15. Women frequently enter prostitution at adolescence, and move to Nevada after the age of 18.

[123] Albert, *Brothel*, page 75.

[124] See 70% reported as a rate of child sexual abuse of those in prostitution by Mimi H. Silbert and Ayala M. Pines (1981) Sexual Child Abuse as an Antecedent to Prostitution. *Child Abuse & Neglect* 5: 407-411; Mimi H. Silbert and Ayala M. Pines (1983) Early Sexual Exploitation as an Influence in Prostitution. *Social Work* 28: 285-289. 74% reported by Evelina Giobbe (1992) Juvenile Prostitution: Profile of Recruitment in Ann W. Burgess (ed.) *Child Trauma: Issues & Research.* New York: Garland Publishing. Evelina Giobbe, Mary Harrigan, Jayme Ryan, and Denise Gamache (1990) *Prostitution: A Matter of Violence against Women.* Minneapolis: WHISPER. 85% reported by Susan Kay Hunter (1994) Prostitution is Cruelty and Abuse to Women and Children. *Michigan Journal of Gender and Law* 1: 1-14.

[125] Melissa Farley et al., Prostitution in Vancouver. As one women explained, "It is internally damaging. You become in your own mind what these people do and say with you. You wonder how could you let yourself do this and why do these people want to do this to you?" Interview with Anonymous Prostituted Woman, in San Francisco, California. May 8, 2004.

[126] Brown, *Sex Slaves*, page 198. Brown discusses psychological conditioning and seasoning by pimps of trafficking victims in Asia.

[127] Albert, *Brothel*, page 234.

[128] Schwartz, *Dialogues with Forgotten Voices*, page 122

[129] Andrea Dworkin (1997) Prostitution and Male Supremacy. In *Life and Death*. New York: Free Press. Also available at http://www.prostitutionresearch.com/c-how-prostitution-works.html.

[130] Albert, *Brothel*, page 135.

[131] See Schwartz, *Dialogues with Forgotten Voices*.

[132] Marianne Wood (1995) *Just a Prostitute*. Queensland: University of Queensland Press, page 55. Author's italics.

[133] Albert, *Brothel,* page 87.

[134] Howard Meadow is at the Las Vegas Municipal Court Regional Justice Center, Alternative Sentencing and Education Division.

[135] Personal Communication (2007) with Christine Stark. April 14, 2007.

[136] Marie Vermeiren (2006) Not for Sale. Video Produced by Coalition against Trafficking in Women (CATW) and the European Women's Lobby (EWL). Available at info@catwinternational.org

[137] Personal Communication (2005) with Paddy Lazar, former staff member of Council for Prostitution Alternatives, Portland, Oregon. Phone interview July 1, 2005. Alluding to the brothel as a mental institution, another woman said, "When I leave the brothel I have all these flashbacks and suicidal memories of my grandfather...In the brothel I've created a structured environment for myself, I cannot fall apart there." Patricia A. Murphy (1993) *Making the Connections: Women, Work, and Abuse.* Orlando: Paul M. Deutsch Press, page 9.

[138] Brown, *Sex Slaves,* page 226.

[139] Cathy Zimmerman, Mazeda Hossain, Kate Yun, Brenda Roche, Linda Morison, and Charlotte Watts (2006) Stolen Smiles: a summary report on the physical and psychological health consequences of women and adolescents trafficked in Europe. London School of Hygiene and Tropical Medicine. Retrieved   December   9,   2006   from www.lshtm.ac.uk/hpu/docs/StolenSmiles.pdf.

[140] In *The Nye County Brothel Wars,* page 13, Jeanie Kasindorf described the confusion and suspicion of Nevada law enforcement personnel who encountered personality changes in a young woman whom they hoped would testify against a violently abusive pimp. One day, the young woman changed from a terrified person who begged for help into a cold streetwise young woman who said she chose to prostitute in the brothel for her pimp. Other days, she stared into space, mumbling as if she were drugged.

[141] There was a statistically significant association between the women's endorsement of having 'trouble remembering important parts of a stressful experience from the past' and the total number of johns serviced. (r =.351, p =.023, N = 42).

[142] Another item that was indicative of symptoms of traumatic stress reached statistical significance: 'Avoiding activities or situations because they reminded you of a stressful experience from the past' (r = .404, p = .007, N = 43).

[143] The relationship between a longer time in prostitution and feeling numb when near family and friends was statistically significant, r = .323, p = .034, N = 43. The longer women had been in prostitution, the more likely they were to state that they had difficulty being close to people they loved.

[144] The relationship between a longer time in prostitution and experiencing flashbacks was statistically significant, r =.368, p = .016, N = 42.

[145] "Dennis Hof's primary business innovation at the Bunnyranch has been to hire porn stars as prostitutes and to promote their presence as a Triple X

Fantasy Camp… When a porn star agrees to work at the ranch for a spell, Hof also tries to arrange for her to appear on Howard Stern's show or in an adult magazine such as *Spectator*, thereby promoting her movies and his brand simultaneously. Rebecca Mead (2001) American Pimp. *New Yorker*, April 23, 2001. Retrieved July 9, 2003 from http://www.rebeccamead.com/2001/2001_04_23_art_reno.htm.

[146] The relationship between being coerced by a john to imitate pornography and traumatic flashbacks was statistically significant (r =.328, p = .030, N = 44).

[147] The relationship between having pornography made of one's prostitution and feeling extremely upset when reminded of past traumatic stress was statistically significant (r = .392, p = .009, N = 43).

[148] Sullivan, *Making Sex Work*, page 106.

[149] For a summary see Farley, 'Bad for the Body, Bad for the Heart,' pages 1087-1125.

[150] Anderson, Working in Nevada.

[151] It is extremely difficult to transmit HIV from prostitute to john but not impossible.

[152] Terri Coles (2006) *Multiple partnerships fueling AIDS epidemic*. Reuters U.K. August 15, 2006. Discussing a paper by Daniel Halperin, USAID, Southern Africa, presented at the 16th Global AIDS conference in Toronto, Canada.

[153] Larson and Narain found that the higher the number of johns, and the higher the number of overall sex partners, the higher women's rate of HIV in Cambodia and Thailand. Heidi J. Larson and Jai P. Narain (2001) *Beyond 2000: Responding to HIV/AIDS in the new millennium*. New Delhi: World Health Organization (WHO) Regional Office for South-East Asia. Retrieved November 15, 2005 from http://w3.whosea.org/EN/Section10/Section18/Section356/Section410.htm, page 17.

[154] Lindeman, Angel's Ladies.

[155] James C. McKinley (2005) A New Law in Tijuana Regulates the Oldest Profession. *New York Times* December 13, 2005. Retrieved May 10, 2006 from http://www.nytimes.com/2005/12/13/international/americas/13prostitutes.html?ex=1292130000&en=e2b22abb426fdbe0&ei=5088&partner=rssnyt&emc=rss.

[156] A Nye county brothel used sniffer dogs for random drug searches of women in the brothel. A pimp stated that at his brothel there were 6000 regular johns and that it was impossible to test all the johns for HIV. Instead, he said, the women in the brothel were tested for HIV because they are a smaller and "controlled" population. Report of a panel of brothel owners at the Desiree Alliance Whores Conference, July 2006, University of Nevada at Las Vegas http://www.desireealliance.org/conference.htm. Remarks quoted here are from a blog about the conference retrieved August 25 2006 from http://sexworkerrights.wordpress.com/page/2/.

[157] Nevada brothel lobbyist George Flint reported in 2006 that Nevada conducts 59,000 Elisa tests on women in the Nevada brothels every month and that no one has ever tested positive. Elisa tests for HIV antibodies. In order for the Elisa test to be effective, a three-month waiting period after possible infection is necessary for the antibodies to appear. Report of a panel of brothel owners at the Desiree Alliance Whores Conference, July 2006. The author has additional documentation on file regarding Elisa testing for HIV infection.

[158] Kuo, *Prostitution Policy*, page 130.

[159] James C. McKinley (2005) A New Law in Tijuana Regulates the Oldest Profession. *New York Times* December 13, 2005. Retrieved May 10, 2006 from http://www.nytimes.com/2005/12/13/international/americas/13prostitutes.html?ex=1292130000&en=e2b22abb426fdbe0&ei=5088&partner=rssnyt&emc=rss.

[160] Personal Communication (2006) with Karen Gedney, M.D. Carson City, Nevada. April 18, 2006.

[161] Glen Puit (1996) Police See Sex Trade: HIV-Infected Prostitutes Serve Too Little Time And Then Return To The Streets, Las Vegas detectives say. *Las Vegas Review Journal* November 21, 1996. Retrieved March 8, 2006 from www.lasvegasreviewjournal.com.

[162] Leonard Cler-Cunningham & Christine Christensen (2001) Violence against women in Vancouver's street level sex trade and the police response. Vancouver: PACE Society.

[163] Bebe Loff, Cheryl Overs, and Paulo Longo (2003) Can health programmes lead to mistreatment of sex workers? *Lancet* 36: 1982-3. June 7, 2003.

[164] Vijayendra Rao, Indrani Gupta, Michael Lokshin, Smarajit Jana (2003) Sex Workers and the Cost of Safe Sex: The Compensating Differential for Condom Use in Calcutta. *Journal of Development Economics* 71 (2): 585-603.

[165] Janice Raymond, Donna Hughes, & Carole Gomez (2001) *Sex Trafficking of Women in the United States: Links Between International and Domestic Sex Industries.* N. Amherst, MA: Coalition Against Trafficking in Women. Available at www.catwinternational.org.

[166] See a discussion of johns' refusal to use condoms in Farley, 'Bad for the Body, Bad for the Heart.'

[167] Kuo, *Prostitution Policy*, page 84.

[168] We heard from several informants that the state of Nevada provided free condoms to the brothels but pimps charged the women for the condoms. If true, this policy discourages condom use. This same practice was reported by both pimps and women in prostitution in Sonagachi, a large brothel area in Kolkata, India. Donations of condoms from producers were offered in Indian prostitution zones, but the pimps charged desperately poor women for each condom. The practice was promoted in India under the camouflage of social marketing.

[169] A confidential source told us that a pimp who controlled a legal brothel collaborated with a pharmaceutical company in the unapproved testing of medications on women in his brothel. This occurred in the 1980s and should be further investigated.

[170] Melissa Farley (2003) *Preliminary Report on Prostitution in New Zealand.* Unpublished manuscript on file with author. May 14, 2003.

[171] Albert, *Brothel*, page 158. Albert explained, "Drinking and drugging often helped some of the women cope with anxiety, boredom, and long work hours," page 159. Albert noted, as we have discussed elsewhere, that many of the women began their drug or alcohol abuse *subsequent to* entry into prostitution, page 160.

[172] Barry, *The Prostitution of Sexuality*, page 232.

[173] Mead, American Pimp.

[174] Anderson, in a separate report, reported "rushed, inadequate" medical exams with physicians who were patronizing and sexist. Anderson, Working in Nevada.

[175] Whiteside, The Invisible Woman, page 20.

[176] Albert, *Brothel*, page 48.

[177] Albert, *Brothel*, page 92, 167.

[178] Anderson, Working in Nevada.

Chapter 3. Pimp Subjugation of Women by Mind Control

[179] Kathleen Barry (1995) *The Prostitution of Sexuality: the Global Exploitation of Women.* New York: New York University Press. See especially chapter 6: Pimping: the Oldest Profession. See also Cecilie Hoigard and Liv Finstad (1992) *Backstreets: Prostitution, Money, and Love.* University Park, PA: Pennsylvania State University Press.

[180] Cathy Zimmerman, Mazeda Hossain, Kate Yun, Brenda Roche, Linda Morison, and Charlotte Watts. (2006) Stolen Smiles: a summary report on the physical and psychological health consequences of women and adolescents trafficked in Europe. London School of Hygiene and Tropical Medicine. Retrieved December 9, 2006 from www.lshtm.ac.uk/hpu/docs/StolenSmiles.pdf.

[181] Steven Hassan (1990) *Combating Cult Mind Control.* Rochester, VT: Park Street Press.

[182] Judith Herman (1992) *Trauma and Recovery.* New York: Basic Books, page 91.

[183] See Hassan, *Combating Cult Mind Control;* Harvey L. Schwartz (2000) *Dialogues with Forgotten Voices: Relational Perspectives on Child Abuse Trauma and the Treatment of Severe Dissociative Disorders* (2000) New York: Basic Books; and also W.H. Bowart (1994/1978) Operation Mind Control: Researcher's Edition. Fort Bragg, California: Flatland Editions.

[184] Schwartz, *Dialogues with Forgotten Voices,* page 314.

[185] Raymond Bechard (2006) *Unspeakable The Hidden Truth Behind The World's Fastest Growing Crime*. New York: Compel Publishing. See also Kathryn Farr (2004) *Sex Trafficking: The Global Market in Women and Children*. New York: Worth Publishers.

[186] Andrew Jacobs (2004) Call Girls, Updated. *New York Times* October 12, 2004.   Retrieved   November   30,   2004   from http://www.nytimes.com/2004/10/12/nyregion/12madam.html?ex=1098599 063&ei=1&en=20de55812ce4a4a7.

[187] The young woman in this example was interviewed by Melissa Farley as defense expert in a California case.

[188] Peter Landesman (2004) The Girls Next Door. *New York Times Magazine* January   25,   2004.   Retrieved   February   18,   2004   from http://www.nytimes.com/2004/01/25/magazine/25SEXTRAFFIC.html.   See description of forced drugging used in UK. Cathy Zimmerman, Katharine Yun, Inna Shvab, Charlotte Watts, Luca Trappolin, Mariangela Treppete, Franca Bimbi, Brad Adams, Sae-Tang Jiraporn, Ledia Beci, Marcia Albrech, Julie Bindel and Linda Regan (2003) Health Risks and Consequences of Trafficking in Women and Adolescents. London School of Tropical Hygiene and Medicine. Retrieved June 19, 2006 from http://www.lshtm.ac.uk/hpu/new_papers.htm.

[189] There is a similarity between fascistic cult leaders who promote pregnancy as a way of increasing membership (the Fundamentalist Latter Day Saints) and pimps' promotion of pregnancy for essentially the same reasons. Both groups of perpetrators breed women for economic gain.

[190] Location where stolen vehicles are stripped down and their parts are 'chopped up' and sold separately.

[191] Michael D. Langone (1993) *Recovery From Cults: Help For Victims of Psychological and Spiritual Abuse*. New York: Norton.

[192] Anna C. Salter (1995) Transforming Trauma: A Guide to Understanding and Treating Adult Survivors of Child Sexual Abuse. Thousand Oaks: Sage Publications.

[193] Adam Hochchild (1998) *King Leopold's Ghost: a Story of Greed, Terror, and Heroism in Colonial Africa*. New York: Houghton Mifflin. See especially pgs 65, 175.

Chapter 4. Johns in Legal and Illegal Prostitution

[194] Gabriel R. Vogliotti (1975) *The Girls of Nevada*. Secaucus NJ: Citadel Press, page 235.

[195] Louise Brown (2000) *Sex Slaves: the Trafficking of Women in Asia*. London: Virago, page 14.

[196] See a further discussion of Grace Quek, who was prostituted by 250 johns in a gang rape that was filmed in Melissa Farley (2006) Prostitution, Trafficking,

and Cultural Amnesia: What We Must *Not* Know in Order To Keep the Business of Sexual Exploitation Running Smoothly. *Yale Journal of Law and Feminism* 18:109-144, pages 134-136.

[197] Personal communication (2006) with Twiss Butler. July 28, 2006.

[198] Kathryn Hausbeck and Barbara G. Brents (2000) Inside Nevada's Brothel Industry, page 230. In Ronald Weitzer (ed.) *Sex for Sale: Prostitution, Pornography and the Sex Industry.* New York: Routledge, pages 217-241. See page 221

[199] Douglas Lindeman (2000) *Angel's Ladies.* Quoting Mack. Lindeman produced and directed the documentary film. Berkeley: Film Kitchen. For purchase contact Orly Ravid at orlyravid@comcast.net.

[200] Brown, *Sex Slaves,* page 131.

[201] Brent K. Jordan (2004) *Stripped: Twenty Years of Secrets From Inside the Strip Club.* Kearny Nebraska: Satsyu Multimedia Press, page 154.

[202] Alexa Albert (2001) *Brothel: Mustang Ranch and Its Women.* New York: Random House, page 182.

[203] E. McLeod (1982) *Women Working: Prostitution Now.* London: Croom Helm, page 63.

[204] Meredith May (2006) Sex Trafficking: San Francisco Is A Major Center For International Crime Networks That Smuggle And Enslave. *San Francisco Chronicle* October 6, 2006. Retrieved October 28, 2006 from http://www.sfgate.com/cgibin/article.cgi?f=/c/a/2006/10/06/MNGR1LGU Q41.DTL&hw=meredith+may+sex+slaves&sn=005&sc=497.

[205] Susan Meiselas (2003, Second Edition, Revised) *Carnival Strippers.* New York: Whitney Museum of American Art, page 50.

[206] Julie O'Connell Davison (1998) *Prostitution, Power, and Freedom.* University of Michigan Press: Ann Arbor, page 209.

[207] We have interviewed men in India, Spain, Scotland and United States, and will be including men from Cambodia in 2008. Two publications contain additional information about preliminary findings of this research. Melissa Farley (2007) 'Renting an Organ for 10 Minutes:' What TricksTell us about Prostitution, Pornography, and Trafficking. In D. Guinn (ed.) Pornography: Driving the Demand for International Sex Trafficking. Los Angeles: Captive Daughers Media. Also in Melissa Farley (2006) Prostitution, Trafficking, and Cultural Amnesia: What We Must *Not* Know in Order To Keep the Business of Sexual Exploitation Running Smoothly. *Yale Journal of Law and Feminism* 18:109-144.

[208] Melissa Farley (2006) Prostitution, Trafficking, and Cultural Amnesia: What We Must Not Know in Order To Keep the Business of Sexual Exploitation Running Smoothly. *Yale Journal of Law and Feminism* 18:109-144.

[209] Jordan, *Stripped,* page 25. Jordan also commented on page 28, "A topless entertainer who has been in the business for more than a week has heard every line, been propositioned from every angle, been lied to, promised, begged and coerced for sex in every manner imaginable."

210 Sarah Katherine Lewis (2006) *Indecent: How I Make it and Fake it as a Girl for Hire.* Emeryville, California: Seal Press, page 52.

211 Erik Hedegard (2007) The Girlfriend Experience. *Playboy.* July 2007, page 72.

212 See Mary Lucille Sullivan (2007) *Making Sex Work: A Failed Experiment with Legalized Prostitution.* North Melbourne:Spinifex. See also Richard J. Estes and Neil A. Weiner (2001) Commercial Sexual Exploitation of Children in the U.S., Canada, and Mexico. University of Pennsylvania.

213 See Kathryn Farr (2004) Militarized Rape and Other Patriarchal Hostilities: Fueling and Legitimating Male Demand for a Sex Trade and also her chapter titled The Organization of Military Prostitution in Modern Times: Building a Sex Trade from Militarized Demand. In Kathryn Farr (2004) *Sex Trafficking: The Global Market in Women and Children.* New York: Worth Publishers.

214 Albert, *Brothel,* page 96.

215 A message posted by a john described his plans for prostitution tourism in Nevada and Costa Rica. "Nevada and Costa Rica trip June 29, 2005, 05:57:45 AM I've been quiet on the boards, but have been planning my annual trip. this time I'll be able to do a hike in the Ruby Mts. near Elko as well as check out the brothels in Carlin, Elko and Wells. I'm not heading home to Madison, but Milwaukee, where I leave for Costa Rica on the 15th, spending four nights in San Jose. This is where I expect to do most of my partying. I'll see what I can get in Nevada, but I'm holding the line on price this time. When I return from CR, it's off to Vegas--but I'm not staying there. It will be a combo of Pahrump and Beatty so I can do hikes in Red Rock and Mt. Charleston and stop at the SLR and Angel's, along with having fun getting walked at Sheri's Tom" http://www.nevadabrothelgirls.com/smf/index.php?topic=5276.0. Pahrump is the location of several brothels, one of which is Sheri's.

216 Debating the economics of prostitution in Nevada, as opposed to Tijuana Mexico, johns debate price versus transportation, even taking a poll that demonstrated that johns prefer a cheaper price for prostitution, traveling to wherever they can obtain women and girls cheaply. http://nvbrothels.net/forum/showthread.php?p=341256. " Last month I visited [Tijuana] and checked out ...Bar and found out a top-earner there of comparable beauty would run $70 with a room for a half-hour. ...club girls at least have comparable health status with girls [from Nevada brothels], at least enough to minimize the risks....Doing the math a two hour session with a girl at Adelita's would run me $280 plus another $400 for transportation if I went cheaply with a layover, pun not intended.. Other countries and options are cheaper still. Now why does the [Nevada legal sex] industry not cater to those of less means? I mean have any brothels where one can get their needs met for lets say $100 for a half-hour." http://www.nevadabrothelgirls.com/smf/index.php?topic=5276.0.

217 "The best bet in Vegas is to cruise the hotel lobbies and bars where many girls are looking for customers -- they expect you're in the hotel and can take

them upstairs. This usually works best if you're in the hotel. I've seen it done a lot at the Hilton, and MGM Grand late at night. But I've had good success with the services that advertise in those "give-away" papers -- they call it exotic dancing but there's no dancing. It's straight sex. Some of the girls who work for these services are the same ones who cruise the casinos while they're waiting for a call." Finally, why not just try at your hotel???? Sitting at the bar in any of the Strip hotels after 10:00 pm or so, will almost always lead to a conversation with a willing lady. You can usually spot the hookers because they will be better dressed (and better looking) then most of the guests. You can usually get one up to your room for a quickie for about $200. Only problem is that most are only willing to go with guests who are staying at the hotel they are working, so choice can be limited.

http://www.worldsexguide.org/las-vegas.txt.html.

[218] Howard Meadow's program for men arrested for soliciting prostitution is a one day program that costs $450, and if the program is completed the charge is changed to jaywalking. No repeat offenders and no registered sex offenders are permitted to participate in the program. From 1998 to 2005, 1549 men completed the diversion program, with 33% of the class monolingual Spanish speakers. 33% of the population of Las Vegas is Latino.

[219] This is one pimp's estimate, and may not reflect the demographics of johns in other brothels.

[220] Jordan, *Stripped*, 2004, pages 132-140.

[221] G. Rivlin (2007) Las Vegas caters to high rollers. *New York Times* June 13, 2007.
http://www.nytimes.com/2007/06/13/business/13vegas.html?_r=1&oref=slo gin

[222] Brown, *Sex Slaves*, pages 137-8.

[223] Albert, *Brothel*, page 234. Another woman - in Douglas Lindeman's documentary *Angel's Ladies* - struggled to describe what it was like prostituting in a Nevada brothel: "This is so unreal, in the middle of the desert, catering mainly to truckers. Every day you dress up, act like a Barbie doll and get paid for it. Sometime it's repulsive. The johns have such unexciting, boring lives. I think of political things, the injustices of society, and how people get away with things in other walks of life."

[224] Albert, *Brothel*, page 19.

[225] Suki Falconberg (2007) In the Las Vegas Sex Industry, the Rapists Go Free. Retrieved May 28, 2007 from
http://www.americanchronicle.com/articles/viewArticle.asp?articleID=27149. May 15, 2007.

Chapter 5. Trafficking for Legal and Illegal Prostitution in Nevada

[226] Meredith May (2006) San Francisco Is A Major Center For International Crime Networks That Smuggle And Enslave. *San Francisco Chronicle* October 6, 2006. Retrieved October 11, 2006 from http://www.sfgate.com/cgibin/article.cgi?f=/c/a/2006/10/06/MNGR1LGU Q41.DTL&hw=meredith+may+sex+slaves&sn=005&sc=497. Debt bondage means that the woman who is prostituted is expected to pay off huge and sometimes escalating debt to 'repay' traffickers and pimps for travel, lodging, food, etc. The terms of debt bondage often are so impossible that women simply decide to escape with nothing at all after years of trying to pay off debts via servicing frighteningly large numbers of johns.

[227] Mary Lucille Sullivan (2007) *Making Sex Work: A Failed Experiment with Legalized Prostitution.* Spinifex: North Melbourne, page 229.

[228] Peter Landesman (2004) The Girls Next Door. *New York Times Magazine* January 25, 2004. Retrieved February 18, 2004 from http://www.nytimes.com/2004/01/25/magazine/25SEXTRAFFIC.html.

[229] See United Nations, Commission on Human Rights, Report of the Special Rapporteur on the Human Rights Aspects of the Victims of Trafficking in Persons, Especially Women and Children, 9 U.N. Doc. E/CN.4/2006/62 (Feb. 20, 2006). See also Janice G. Raymond (2003) Ten Reasons for *Not* Legalizing Prostitution and a Response to the Demand for Prostitution. In Melissa Farley (ed.) (2003) *Prostitution, Trafficking, and Traumatic Stress.* Binghamton: Haworth, pages 315-332. Also available at Coalition Against Trafficking in Women website. http://www.catwinternational.org/.

[230] Janice G. Raymond (2003) Ten Reasons for *Not* legalizing Prostitution and a Response to the Demand for Prostitution. In M. Farley (ed.) (2003) *Prostitution, Trafficking, and Traumatic Stress.* Binghamton: Haworth, pages 315-332. Available at http://www.catwinternational.org/.

[231] John L. Smith (2001) Las Vegas has a long way to go in dealing with sexual harassment. *Las Vegas Review-Journal* April 25, 2001. Retrieved May 22, 2005 from http://www.reviewjournal.com/lvrj_home/2001/Apr-25-Wed-2001/news/15937128.html.

[232] Las Vegas is in Clark County and is the state of Nevada's largest urban center. In 2007 the population of Clark County was 1,375,765. City of Las Vegas Clark County website, http://www.lasvegascitywebsite.com/. Nevada state law prohibits legal prostitution in any county with a population over 400,000.

[233] See Steve Miller's series of columns *Inside Vegas* including The Rick Rizzolo Connection at http://www.stevemiller4lasvegas.com/RickRizzolo.html

234 Michael J. Goodman and William C. Rempel (2006) In Las Vegas, They're Playing With a Stacked Judicial Deck: Some judges routinely rule in cases involving friends, former clients and business associates -- and in favor of lawyers who fill their campaign coffers. *Los Angeles Times* June 8, 2006. Retrieved September 8, 2006 from http://www.latimes.com/news/nationworld/nation/la-na vegas8jun08,0,453414,full.story?coll=la-home-headlines.

235 Louise Brown (2000) *Sex Slaves: The Trafficking of Women in Asia.* London: Virago, page 123.

236 Brown, *Sex Slaves,* page 247.

237 Esohe Aghatise (2004) Trafficking for Prostitution in Italy: Possible Effects of Government Proposals for Legalization of Brothels. *Violence Against Women* 10 (10): 1126-1155.

238 Janice G. Raymond, Jean D'Cunha, Siti Ruhaini Dzuhayatin, H. Patricia Hynes, Zoraida Ramirez Rodriguez, and Aida Santos (2002) *A Comparative Study of Women Trafficked in the Migration Proce*ss. Amherst, MA. Coalition Against Trafficking in Women. Raymond and colleagues found that women in domestic United States prostitution reported higher levels of violence than women who had been trafficked into the United States. Retrieved June 7, 2007 from http://action.web.ca/home/catw/readingroom.shtml?x=17062&AA_EX_Sess ion=83b3cce42ff04856676b3a24b959634a.

239 X. Salgado (2002) Victim Assistance & Sexual Assault Program (VASAP) in Montgomery County, Maryland. Presentation at Seminar on Human Trafficking. Jan 11, 2002. Falls Church, Virginia.

240 Melissa Farley, Ann Cotton, Jacquelyn Lynne, Sybile Zumbeck, Frida Spiwak, Maria E. Reyes, Dinorah Alvarez , Ufuk Sezgin (2003) Prostitution and Trafficking in 9 Countries: Update on Violence and Posttraumatic Stress Disorder. *Journal of Trauma Practice* 2 (3/4): 33-74. Also in Melissa Farley (ed.) (2003) *Prostitution, Trafficking, and Traumatic Stress.* Binghamton: Haworth, page 43. 89% is a conservative figure - many service providers report that almost all women want to escape prostitution.

241 John L. Smith (2005) Standalone, strip mall massage parlors scrutinized for prostitution. *Las Vegas Review-Journal* May 11, 2005. Retrieved May 28, 2005 at http://www.reviewjournal.com/lvrj_home/2005/May-11-Wed-2005/news/26491311.html. Smith is sympathetic to those he considers to be trafficking victims, but like many other observers, he fails to see the harm unless the prostitution is happening under extreme physical coercion.

242 Brown, *Sex Slaves,* page 10.

243 Thank you Samantha Berg for this important concept.

244 Thanks to Louise Brown for this notion of a hierarchy in prostitution and some of these concepts. In Brown, *Sex Slaves,* page 17.

245 Ian Urbina (2006) Keeping It Secret As the Family Car Becomes a Home. *New York Times* April 2, 2006. Study conducted by the National Low Income

Housing Coalition. Retrieved April 30, 2006 from http://www.nytimes.com/2006/04/02/us/02cars.html?ex=1144641600&en=3 36986e591dbeef2&ei=5070&emc=eta1.

[246] Taylor Lee (2004) In and Out: a Survivor's Memoir of Stripping. In Christine Stark and Rebecca Whisnant (eds.) *Not for Sale: Feminists Resisting Pornography and Prostitution*. North Melbourne: Spinifex, pages 56-63.

[247] Molly Chattopadhyay, S. Bandyopadhyay, and C. Duttagupta (1994) Biosocial Factors Influencing Women to Become Prostitutes in India. *Social Biology* 41:252.

[248] Juliana Beasley (2003) *Lapdancer*. New York: Powerhouse Books.

[249] Margaret A. Baldwin (2003) Living in Longing: Prostitution, Trauma Recovery, and Public Assistance. In Melissa Farley (ed.) *Prostitution, Trafficking, and Traumatic Stress*. Binghamton: Haworth Press, pages 267-314.

[250] Brown, *Sex Slaves*, page 55.

[251] Brown, *Sex Slaves*, page 57.

[252] Jean Enriquez (2006) *Globalization, Militarism and Sex Trafficking*. Speech at Women World March. July 4 – 9, 2006. Lima, Peru. Retrieved March 23, 2007 from http://www.catw-ap.org/speeches_papers/20061118_gmst.htm.

[253] Monica O'Connor and Grainne Healy (2006) The Links Between Prostitution and Sex Trafficking: a Briefing Handbook. European Women's Lobby and Coalition Against Trafficking in Women. Retrieved March 3, 2007 from http://action.web.ca/home/catw/readingroom.shtml?x=89931&AA_EX_Sess ion=c06a11e1a3d8abd6cd9b8ecc3627ce8b.

[254] Catharine A. MacKinnon (2005) Liberalism and the Death of Feminism. In *Women's Lives, Men's Laws*. Cambridge: Harvard University Press, page 260.

[255] See Catharine A. MacKinnon (1993) *Prostitution and Civil Rights. Michigan Journal of Gender and Law* 1: 13-31. See also Tanya K. Hernandez (2001) Sexual Harassment and Racial Disparity: The Mutual Construction of Gender and Race, U. *Iowa Journal of Gender, Race & Justice* 4:183-224.

[256] Associated Press (2007) Man Who Ran Ring of Brothels Sentenced. Apr. 14, 2007 WRAL.com, Raleigh-Durham-Fayetteville. Retrieved April 15, 2007 from http://www.wral.com/news/national_world/national/story/1273570/.

[257] Brian Haynes (2007) Operation Doll House: Bust 'just tip of the iceberg:' Authorities suspect women forced into prostitution. *Las Vegas Review-Journal* April 24, 2007. Retrieved April 25, 2007 from http://www.lvrj.com/news/7164646.html.

[258] K.C. Howard (2007) FBI says human trafficking not involved in brothel raids. *Las Vegas Review-Journal* May 10, 2007. Retrieved May 10, 2007 from http://www.lvrj.com/news/7436826.html.

[259] Survivor of domestic trafficking speaking at Western Regional Seminar on Human Trafficking (2005) San Diego, California. September 29, 2005.

260 Dot Reidelback and Laurie Allen (2006) Banking on Heaven: Polygamy in the Heartland of the American West. Over the Moon DVD. BankingOnHeaven.com.

261 Janice G. Raymond (2003) Ten Reasons for *Not* legalizing Prostitution and a Response to the Demand for Prostitution. In Melissa Farley (ed.) (2003) *Prostitution, Trafficking, and Traumatic Stress.* Binghamton: Haworth, pages 315-332. Available at http://www.catwinternational.org/.

262 Lenore Kuo (2002) Prostitution Policy: Revolutionizing Practice through a Gendered Perspective. New York: New York University Press, page 80.

263 California, Utah, Florida, New York, Ohio, Arizona, South Carolina, New Jersey were the most commonly reported states from which women were moved to Nevada.

264 Mary Lucille Sullivan (2007) *Making Sex Work: A Failed Experiment with Legalized Prostitution.* Spinifex: North Melbourne, page 243 .

265 Jack Sheehan (2004) *Skin City: Uncovering the Las Vegas Sex Industry.* Las Vegas: Stephens Press, page 188. Sheehan's information is based on interviews with Clark County Sheriff Bill Young and undercover police officers from Las Vegas Metro Police Department.

266 Case Number 06-cr-0014 JSW. Antoine Vernon Mouton, age 24, was indicted on child sex-trafficking charges after he allegedly moved a 13-year-old girl from San Francisco to Las Vegas to prostitute. He allegedly pimped the girl during the month of April and kept all the money she earned. Mouton was arrested in Las Vegas on unrelated burglary and theft charges. The case came to light when Mouton attacked the girl because she had run away from him in Las Vegas after she was arrested on prostitution charges on April 7, 2005. Chronicle Staff Report (2006) SF man accused of pimping 13-year-old girl. *San Francisco Chronicle* January 31, 2006. Retrieved February 2, 2006 from http://www.sfgate.com/cgibin/article.cgi?f=/c/a/2006/01/31/MNGAMH0 KNS7.DTL.

267 Juliet W. Casey (2005) Human traffic targeted: authorities launch second campaign. *Las Vegas Review-Journal* March 16, 2005. Retrieved September 9, 2005 from http://www.reviewjournal.com/lvrj_home/2005/Mar-16-Wed-2005/news/26081825.html.

268 Case Number CR-S-05-0288-RCJ(PAL) United States v. Darwin Hayter. On May 17, 2006, Darwin Hayter pleaded guilty to two counts of Transportation of a Minor for Prostitution for transporting two minors and an adult female from Aurora, Colorado, to Las Vegas, Nevada, and Los Angeles, California, in June 2005 with the intent that they engage in prostitution and illegal sexual activity. Hayter is scheduled to be sentenced on October 16, 2006.

269 Case Number 2:05-cr-00627-SGL United States v Doss et al. On June 26, 2006, a jury in the Central District of California found Juan Rico Doss guilty of two counts of sex trafficking of children, three counts of transporting minors

into prostitution, one count of conspiracy to commit sex trafficking of children and transporting minors into prostitution, and two counts of witness tampering.   Retrieved   November   18,   2006   from http://releases.usnewswire.com/GetRelease.asp?id=68378.

[270] Associated Press (2005) Slaying Attributed to Involvement with Rappers or Prostitution.   Retrieved   December   6,   2005   from http://www.lasvegassun.com/sunbin/stories/nevada/2005/dec/05/120510 644.html.

[271] Karrine Steffans (2005) *Confessions of a Video Vixen.* New York: Harper Collins. See also *Hip Hop: Beyond Beats & Rhymes (2006),* a film by Byron Hurt that documents and challenges the misogyny and homophobia of rap music. Media Education Foundation.

[272] FBI Press Release (2005) December 16, 2005. Atlantic City, NJ. The case, which includes money laundering, was a result of joint efforts on the part of the FBI, the Atlantic City, Las Vegas, Pleasantville, and Egg Harbor Police Departments, the New Jersey State Police, New Jersey Division of Gaming Enforcement, the U.S. Department of Labor, Office of Labor Racketeering, and the U.S. Postal Service. Retrieved April 22, 2006 from http://www.fbi.gov/dojpressrel/pressrel05/innocencelost.htm.

[273] A man with a prior conviction for pandering a child stemming from his transportation of other minors for the purpose of prostitution, was found guilty in this case of two counts of sex trafficking of children, three counts of transporting minors into prostitution, one count of conspiracy to commit sex trafficking of children and transporting minors into prostitution, and two counts of witness tampering. Newswire, Justice Department. (2006) Nevada Man Convicted on Charges Related to the Sex Trafficking of Minors. Retrieved June 27, 2006 from http://releases.usnewswire.com/GetRelease.asp?id=68378.

[274] Pimps and traffickers use methods of brainwashing and programming that have been developed by political torturers. For more information on this process see Harvey L. Schwartz (2000) *Dialogues with Forgotten Voices: Relational Perspective on Child Abuse Trauma and Treatment of Dissociative Disorders.* New York: Basic Books. See also Judith Herman's essential reading on this topic, *Trauma and Recovery* (1992) New York: Basic Books.

[275] Survivor of domestic trafficking speaking at Western Regional Seminar on Human Trafficking in San Diego, September 29, 2005.

[276] Case Number CR-S-05-0137-RCJ United States v. Louis Wright. On January 9, 2006, Louis Kenneth Wright aka Kenny Red aka Tyler Montana, a nationally-known pimp who has appeared in documentary films about pimping, was sentenced to 30 months in prison and ordered to pay a $100,000 fine for his guilty pleas to two counts of money laundering. Wright laundered the proceeds of his prostitution business and identity theft crimes through the purchase of two properties in Las Vegas.

---

[277] Case Number CR-S-03-0100-RLH(LRL) Jonathan Duke Flake et al. On August 15, 2005, Jonathan Duke Flake, aka Caribbean, aka Corey Robinson, aka Brandon Cashman, was sentenced to 77 months in prison for his guilty plea to money laundering. Flake recruited young women in Portland, Oregon, and pimped them in Las Vegas, Nevada. Flake directed the prostitutes to deposit all monies they earned from prostitution activities into a bank account owned by his co-defendant, Portland Oregon attorney Cheryl Chadwick. Chadwick also made rent, automobile, and attorney fee payments for Flake with the monies that Flake or others deposited into the account. Chadwick pleaded guilty to Money Laundering and was sentenced to five years probation. Two prostitutes also pleaded guilty to Money Laundering for their roles in the offense and were sentenced to three years probation. Case Number CR-S-03-0046-KJD(RJJ) United States v. Quinton Williams. In 2003, a trafficker was convicted of money laundering, transporting a minor for prostitution, transporting an adult for prostitution, sex trafficking of children, and interstate travel in aid of racketeering. Williams was sentenced to 125 months in prison for operating an interstate prostitution business in which he transported women, including minors, from Chicago to Portage, Indiana; Houston, Texas; Phoenix, Arizona; and Las Vegas, Nevada for prostitution. Williams supervised their prostitution activities and kept all of their earnings. Williams has prior felony convictions for attempted robbery and narcotics trafficking.

Chapter 6. Illegal Escort and Strip Club Prostititution in Las Vegas

[278] Joe Schoenmann (2003) Reviving Downtown on the backs of taxed and regulated prostitutes. *Las Vegas Weekly* July 10, 2003. Retrieved November 12, 2005 from http://www.lasvegasweekly.com/2003/07_10/news_coverstory.html.

[279] Jacqueline Lewis (1998) Lap Dancing: Personal and Legal Implications for Exotic Dancers. In James E. Elias, Vern L. Bullough, Veronica Elias, & Gwen Brewer (eds.) *Prostitution: On Whores, Hustlers, and Johns.* Amherst, NY: Prometheus Books, pages 376-389.

[280] Kelly Holsopple (1998) Strip Club Testimony. Retrieved June 28, 2006 from http://www.ccv.org/downloads/pdf/Strip_club_study.pdf.

[281] K. Son (2003) Strip Club Dancer is All Worn Out. *BackPage*. Retrieved September 5, 2003 from http://www.news24.com/News24/Backpage/HotGossip/0,,2-1343 1344_1412175,00.html.

[282] Carol Rambo Ronai and Carolyn Ellis (1989) Turn-Ons for Money: Interactional Strategies of the Table Dancer. *Journal of Contemporary Ethnography* 18: 271-298, page 292.

[283] Ronai and Rambo, Turn-Ons for Money, page 287.

284 Sarah Katharine Lewis (2006) *Indecent: How I Make it and Fake it as a Girl for Hire.* Emeryville: Seal Press, page 195

285 David Strow (2001) Slot machine regulators concerned about LV topless club *Las Vegas Sun* July 13, 2001. Retrieved November 29, 2005 from http://www.casinowatch.org/sex_prostitution/sexessful_gambling.html.

286 A john wrote about prostitution in the Las Vegas strip clubs: "There are a number of strip clubs you can go to, and some of the girls will meet you after work, but here again cost can be a problem. Unless somebody tips you off to a particular lady that is interested in dates, you can blow a lot of $$ and never get anywhere. IF you do find a girl in one of the better clubs, expect to pay her $300 and up. The best Clubs in Vegas are Cheetah's, Olympic Gardens, Crazy Horse 2, and Club Paradise. My personal favorite is Cheetah's. Retrieved September 6, 2006 from http://www.worldsexguide.org/las-vegas.txt.html.

287 See exhaustive documentation of organized crime involvement and control in the Las Vegas strip club Crazy Horse Too at Steve Miller's website listing dozens of linked articles http://www.stevemiller4lasvegas.com/RickRizzolo.html.

288 Every strip club in Las Vegas has lap dances and/or private booths for dancing and prostitution.

289 Hilary Frey (2002) The Naked Truth. *The Nation* November 25, 2002. Retrieved March 7, 2004 from http://www.thenation.com/doc/20021125/frey.

290 Lily Burana (2001) *Strip City*. New York: Hyperion, page 67.

291 Burana, *Strip City*, page 299.

292 Brent K. Jordan (2004) *Stripped: Twenty Years of Secrets From Inside the Strip Club*. Kearny Nebraska: Satsyu Multimedia Press, page 19.

293 Doug McMurdo (2006) Chicken Ranch brothel sells for $5 million. *Pahrump Valley Times* Februrary 8, 2006. Retrieved February 9, 2006 from http://www.pahrumpvalleytimes.com/2006/02/08/news/ranch.html.

294 Joel Stein (2004) The Strip is Back. *Time Magazine* July 26, 2004.

295 Statement of Ken Green, owner of the Chicken Ranch Brothel in Pahrump. Steve Miller (2003) Mayor Wants Brothels Downtown http://www.americanmafia.com/inside_vegas/10-27-03_Inside_Vegas.html. Miller is quoting from an article by Erin Neff and Michael Squires (2003) Councilwoman says she looks forward to resolving matter Sunday. *Las Vegas Review-Journal* October 26, 2003.

296 Personal Communication (2007) with Steve Miller. March 8 2007.

297 Jack Sheehan (2004) *Skin City: Uncovering the Las Vegas Sex Industry*. Las Vegas: Stephens Press, page 24.

298 Sheehan, *Skin City*, page 25.

299 Sheehan, *Skin City*, page 25.

300 Sheehan, *Skin City*, page 29.

301 Sheehan, *Skin City*, page 29.

[302] An investigator noted that there is major theft of liquor from California shipping ports by organized crime. Organized criminals sell much of this stolen liquor to Las Vegas strip clubs, thus the actual cost of liquor to strip clubs in Las Vegas is likely to be very low and the profits large.

[303] These figures are mostly from Brent K. Jordan, a bouncer at Cheetah's. See Jordan, *Stripped* pages 54-59. Both Sheehan (*Skin City*) and Jordan describe in detail the large amounts of cash generated in strip clubs which is moved to many locations and is sometimes paid as managers' cash salaries.

[304] Steve Miller (2006) See a compendium of articles written by Miller about Rick Rizzolo and generally about organized criminal involvement at the Crazy Horse Too strip club in Las Vegas. Retrieved March 28, 2007 from http://www.stevemiller4lasvegas.com/RickRizzolo.html.

[305] Rizzolo also contributed $40,000 to the political campaign of Las Vegas Mayor Oscar Goodman. See sources cited in http://www.twistedbadge.com/story_articles/id_242/. See also Glenn Puit (2005) Federal Investigation: Club manager arrested - Indictment alleges racketeering at Crazy Horse Too. *Las Vegas Review-Journal* January 20, 2005. Retrieved June 20, 2006 from
http://www.reviewjournal.com/lvrj_home/2005/Jan-20-Thu-2005/news/25700796.html.

[306] Faraci will serve 8 to 16 months as part of an agreement by which he pleaded not guilty to tax charges but enables him to avoid admitting he's associated with organized crime. John L. Smith (2006) Deals cut when one corrupt official rats out others, and plea-bargains: government offers Rizzolo a sweet deal he shouldn't refuse. *Las Vegas Review-Journal* May 07, 2006. Retrieved May 10, 2006 from
http://www.reviewjournal.com/lvrj_home/2006/May-07-Sun-2006/news/7258133.html.

[307] Mary Lucille Sullivan (2007) *Making Sex Work: A Failed Experiment with Legalized Prostitution.* Spinifex: North Melbourne, pages 198-199.

[308] Sullivan, *Making Sex Work,* page 189. Australia, like Nevada, has legal prostitution.

[309] Jebbie Whiteside (2004) The Invisible Woman: Domestic Violence and Prostitution. Unpublished paper on file wih author, pages 2-3.

[310] A tourist's neck was broken by a bouncer at the Crazy Horse, resulting in quadriplegia. Steve Miller (2004) Dateline NBC Finally Tells of Brutality at Crazy Horse Too: Vegas Mayor & DA's Inaction Inspires National TV Exposé. Retrieved November 28, 2005 from http://www.americanmafia.com/inside_vegas/8-9-04_Inside_Vegas. See Steve Miller's reporting about violence at the Crazy Horse Too strip club in Las Vegas, including owner Rizzolo's felony conviction for beating a patron with a baseball bat and many other assault and robbery complaints at http://www.stevemiller4lasvegas.com/RickRizzolo.html. See also Brent K.

Jordan (2004) *Stripped: Twenty Years of Secrets From Inside the Strip Club*. Kearny Nebraska: Satsyu Multimedia Press, page 168.

[311] Andreas G. Philaretou (2006) Female Exotic Dancers: Intrapersonal and Interpersonal Perspectives. *Sexual Addiction & Compulsivity* 13: 41-52, page 43.

[312] Lily Burana (2001) *Strip City*. New York: Hyperion, pages 107, 109.

[313] Philaretou, *Female Exotic Dancers*, page 40.

[314] Rebecca Mead (2001) American Pimp. *New Yorker* April 23, 2001. Retrieved January 20, 2006 from http://www.rebeccamead.com/2001/2001_04_23_art_reno.htm.

[315] "What is it like to go home after spending the night bouncing your crotch over the faces of people you don't know? How long does it take to settle back into your body, because you'd have to go pretty far away in order to be that exposed for that long wouldn't you?" Burana, *Strip City*, page 162.

[316] Burana, *Strip City*, page 221.

[317] David Scott (1996) Behind the G-String: an exploration of the stripper's image, her person and her meaning. Jefferson, NC: McFarland.

[318] Jordan, *Stripped*, page 72.

[319] Launce Rake (2005) *Strip Clubs Scrutinized*. *Las Vegas Sun* November 6, 2005. Retrieved December 1, 2005 from http://www.lasvegassun.com/sunbin/stories/sun/2005/nov/06/519618491.html?strip%20clubs%20scrutinized%202005.

7. Domestic and International trafficking for Prostitution and Organized Crime in Nevada

[320] Jen Lawson (2003) Fed Official Urges Battle Against Child Prostitution, *Las Vegas Sun* December 12, 2003. Retrieved February 12, 2006 from http://www.lasvegassun.com/sunbin/stories/sun/2003/dec/12/516011564.html?Fed%20.

[321] Kathryn Hausbeck (2004) Sex Industry and Sex Workers in Nevada. Retrieved February 17, 2006 from http://www.unlv.edu/centers/cdclv/healthnv/sexindustry.html.

[322] Juliet V. Casey (2005) Human traffic targeted: Authorities launch second campaign. *Las Vegas Review-Journal* Wednesday, March 16, 2005. Retrieved March 27, 2005 from http://www.reviewjournal.com/lvrj_home/2005/Mar-16-Wed-2005/news/26081825.html.

[323] Kathryn Farr (2005) *Sex Trafficking: The Global Market in Women and Children*. New York: Worth Publishers, pages 56-57.

[324] Dutch News.nl (2007) Police bust major women trafficking gang. Retrieved February 12, 2007 from http://www.dutchnews.nl/news/archives/2007/02/police_bust_major_women_traffi.php.

[325] Parliamentary Joint Committee on the Australian Crime Commission (2007) Inquiry into the future impact of serious and organized crime on Australian society. Public hearing in Melbourne, Australia May 5, 2007. Retrieved May 8, 2007 from http://www.aph.gov.au/senate/committee/acc_ctte/organised_c rime/hearings/index.htm.

[326] Julie Bindel (2004) Streets Apart. *The Guardian*, UK. May 15, 2004. Available at http://www.prostitutionresearch.com/c-laws-about-prostitution.html.

[327] Janice G. Raymond (2003) Ten Reasons for Not Legalizing Prostitution and a Legal Response to the Demand for Prostitution. In Melissa Farley (ed.) *Prostitution, Trafficking, and Traumatic Stress*, page 315. Sources include the Budapest Group's 1999 Report, and an International Organization of Migration (IOM) 1995 Report. A 2007 study noted an increase in Eastern European women in both legal and illegal Dutch prostitution. This was explained in part by the ease of movement of women European Union countries' borders and the presence of legal prostitution. Daalder, A.L. (2007) Prostitutie in Nederland na opheffing van het bordeelverbod. http://www.wodc.nl/Onderzoeken/Overkoepelend_rapport_evaluatie_opheffi ng_bordeelverbod__1204D_.asp.

[328] Dutch News.nl (2006) Nigerian girls trafficked. Retrieved October 12, 2006 from http://www.dutchnews.nl/news/archives/2006/10/nigerian_girls_trafficked.p hp.

[329] Molly Moore (2007) Changing Patterns in Social Fabric Test Netherlands' Liberal Identity. *Washington Post* June 23, 2007. Retrieved June 23, 2007 from http://www.washingtonpost.com/wp dyn/content/article/2007/06/22/AR2007062202015.html.

[330] Mary Lucille Sullivan (2007) *Making Sex Work: A Failed Experiment with Legalized Prostitution*. Spinifex: North Melbourne, page 207.

[331] Dutch News.nl (2007) Police bust major women trafficking gang. Retrieved February 12, 2007 from http://www.dutchnews.nl/news/archives/2007/02/police_bust_major_wome n_traffi.php.

[332] Lenore Kuo (2002) *Prostitution Policy: Revolutionizing Practice through a Gendered Perspective*. New York: New York University Press, page 86.

[333] Sullivan, *Making Sex Work*, page 243.

[334] Sullivan, *Making Sex Work*, page 186.

[335] Although prostitution is called "sex tourism" for visitors to Las Vegas, and sometimes it's even called simply "sex" for sale on the world wide web, in fact, the sex of prostitution is more accurately described as sex for the john. Prostitution is not sex for the women, who have consistently made that extremely clear. Some women have described it as business, others as rape, but

no women in prostitution claim the john's sex in prostitution to be her own sexuality.

[336] Sarah Shannon (1999) Prostitution and the Mafia: The Involvement of Organized Crime in the Global Sex Trade. In Phil Williams (ed.) *Illegal Immigration and Commercial Sex: The New Slave Trade*. Portland: Frank Cass, page 126.

[337] Sullivan, *Making Sex Work*, page 147.

[338] Jeff German (2002) Inside the FBI's Operation Thin Crust. *Las Vegas Sun*. Retrieved July 22, 2006 from http://www.casinowatch.org/sex_prostitution/sexessful_gambling.html.

[339] Jim Brunner (2007) Strip-club owners charged over campaign contributions. *Seattle Times* June 9, 2007. Retrieved June 8, 2007 from http://seattletimes.nwsource.com/html/localnews/2003737549_colacurcio07 m.html.

[340] Sullivan, *Making Sex Work*, page 205-6. See also Keith Moor (2007) Business washed drug cash. *Herald-Sun*, Australia April 30, 2007. Retrieved May 1, 2007 from http://www.news.com.au/heraldsun/story/0,21985,21641647-2862,00.html?from=public_rss.

[341] Renee Viellaris (2007) Tax Blitz on Sex Industry. January 1, 2007. Retrieved March 2, 2007 from http://www.news.com.au/couriermail/story/0,23739,20998647-952,00.html

[342] Personal Communication (2007) with Gunilla S. Ekberg. July 4, 2007. From 2001 to 2006 Gunilla S. Ekberg was the Special Advisor to the Swedish government on prostitution and trafficking in human beings. She is currently Co-Executive Director of the Coalition Against Trafficking in Women International.

[343] Jeff German (2004) Vegas has new crime element: Israeli mob. *Las Vegas Sun* April 30, 2004. Retrieved April 13, 2005 from http://www.lasvegassun.com/sunbin/stories/commentary/2004/apr/30/5167 82598.html.

[344] German, Vegas has new crime element: Israeli mob.

[345] Scott Deitche (2000) The Gambinos Come to Vegas. *AmericanMafia.com*. Retrieved August 7, 2005 from http://www.americanmafia.com/Feature_Articles_24.html.

[346] See Donna M. Hughes and Tatyana A. Denisova (2001) The Transnational Political Criminal Nexus of Trafficking in Women from Ukraine. *Trends in Organized Crime* 6(3-4). Spring/Summer 2001.

[347] John L. Smith (2005) Del Mar's demise won't dent increasingly sophisticated local sex trade. *Las Vegas Review-Journal* May 8, 2005. Retrieved February 7, 2006 from http://www.reviewjournal.com/lvrj_home/2005/May-08-Sun-2005/news/26464848.html. Case # 05CR 503 Judge SAS.

[348] The case resulted from a joint investigation by U.S. Immigration and Customs Enforcement, the Federal Bureau of Investigation, the Office of

Inspector General for the United States Department of Labor, and the Los Angeles Police Department. FBI, Los Angeles (2005) Los Angeles woman pleads guilty to human trafficking prostitute. December 6, 2005. http://losangeles.fbi.gov/dojpressrel/pressrel05/la120605usa.htm.

349 Juliet Casey (2005) Human traffic targeted: authorities launch second campaign. *Las Vegas Review-Journal* March 16, 2005. Retrieved December 21, 2006 from
http://www.reviewjournal.com/lvrj_home/2005/Mar-16-Wed-2005/news/26081825.html.

350 "There is [a massage parlor] on Spring Mountain Road called China Sauna staffed by mid 30 ish Korean women. For $100 you get an hour divided up into sauna, hand bath, massage and hand job (you may have to ask, $20 tip expected)." http://www.worldsexguide.org/las-vegas.txt.html.

351 Jaxon Van Derbeken and Ryan Kim (2005) Alleged sex-trade ring broken up in Bay Area: Police say Koreans in massage parlors were smuggled in. *San Francisco Chronicle* July 2, 2005. Case number CR05-00395-JSW.

352 Juliet W. Casey (2005) Human traffic targeted: authorities launch second campaign. *Las Vegas Review-Journal* Wednesday, March 16, 2005. Retrieved September 8, 2006 from
http://www.reviewjournal.com/lvrj_home/2005/Mar-16-Wed-2005/news/26081825.html.
Case number CR-S-00-0298-KJD(LRL) United States v. Stanley Chan, et al. On December 21, 2001, Stanley Chan, Dat Ming Leung, Yuk Ching Liu, Ru Xiang Zhao, and Dan Chau were sentenced on their convictions for conspiracy to commit money laundering. The defendants were involved in an international conspiracy in which they illegally smuggled Asian women into the United States and used them as prostitutes in brothels in Nevada, California, and other states. The Asian women were expected to pay the conspirators a fee for arranging their illegal entry into the United States. The women paid off the debt bondage by prostituting at brothels operated by the traffickers. Typically the women prostituted two to three weeks at a brothel, and were then transported to a new brothel, usually in a different state. Chan, Leung, and Liu operated the brothels in Las Vegas; Zhao operated an associated brothel; and Chau provided security for the brothels. Stanley Chan was sentenced to 52 months imprisonment; Dat Ming Leung was sentenced to 48 months imprisonment; Yuk Ching Liu was sentenced to 41 months imprisonment; Ru Xiang Zhao was sentenced to 30 months imprisonment; and Dan Chau was sentenced to 27 months imprisonment. Later, four other defendants involved in the conspiracy also entered guilty pleas to conspiracy to commit money laundering. Douglas Vinh Chau was sentenced to 30 months in prison; Tjui Ha was sentenced to 41 months in prison; Minh Buu Huynh was sentenced to 30 months in prison; and Cindy Tan was sentenced to 27 months in prison. The case arose from the FBI's and Las Vegas Metropolitan Police Department's Asian organized crime

program. It represented the first significant prosecution of human trafficking and Asian organized crime elements in Las Vegas. See also KLAS-TV (2002) Conference Focuses on Dealing With Growing Asian Gangs. April 5, 2005. http://www.klastv.com/Global/story.asp?S=291777&nav=menu102_6.

[353] Marisa Bava Ugarte, Laura Zarate, and Melissa Farley (2003) Prostitution and Trafficking of Women and Children from Mexico to the United States. *Journal of Trauma Practice* 2(3/4). Also available at http://www.prostitutionresearch.com/c-trafficking.html

[354] Annie Fukushima (2007) in a doctoral program in Women's Studies and Ethnic Studies at University of California, Berkeley, contributed this information from her research.

[355] Annie Fukushima (2006) Visiting Las Vegas: Bodies Imagined: Race, Gender and Sexual Difference in Sex Industry Advertisements. Unpublished paper.

[356] Glen Meek (2006) Channel 13 KTNV-TV Report on Brothel Raid, Las Vegas July 3 2006. See also Meek's July 16, 2006 Investigative Report for KTNV-TV.

[357] Louise Brown (2000) *Sex Slaves: the Trafficking of Women in Asia*. London: Virago, pages 107-109.

[358] These countries were named by a law enforcement source as origins of women trafficked to Las Vegas using this particular legitimate business scheme.

[359] Edward J. McDermott (2006) *Journal of Gang Research* 13(2): Winter 2006, pages 27-36.

[360] Personal Communication (2006) with Philip Williams, Professor of International Affairs, Public and Urban Affairs at the Ridgeway Center for International and Security Studies, University of Pittsburg. August 31 2006.

[361] Frida Berrigan (2001) Indonesia at the Crossroads: U.S. Weapons Sales and Military Training: A Special Report. *Arms Trade Resource Center*. October 2001. Retrieved June 7, 2007 from http://www.worldpolicy.org/projects/arms/reports/indo101001.htm.

[362] Lesley McCulloch (2000) Business as Usual. *Inside Indonesia* July-September 2000. Cited in Frida Berrigan (2001) Indonesia at the Crossroads: U.S. Weapons Sales and Military Training: A Special Report. *Arms Trade Resource Center*. October 2001 Retrieved June 25, 2007 from http://www.worldpolicy.org/projects/arms/reports/indo101001.htm.

[363] Lesley McCulloch (2003) Greed: the silent force of the conflict in Aceh. University of Deakin, Melbourne, Australia. October 2003. Retrieved June 25, 2007 from http://64.233.179.104/scholar?hl=en&lr=&q=cache:4EhBU0mWF-cJ:www.preventconflict.org/portal/main/greed.pdf+Lesley+McCulloch.

[364] US-Russia Press Club (2003) Mogilevich indicted in Pennsylvania for racketeering, money laundering. Retrieved June 25, 2007 from http://www.moscowtelegraph.com/mogilevich.htm.

365 Small Arms Trade (2006) *Small Arms Trade Transparency Barometer.* Retrieved July 17, 2007 from http://www.smallarmssurvey.org/files/portal/issueareas/transfers/baro.html.

366 Personal Communication (2007) with Kathryn Farr. July 18, 2007.

367 See Farr's Militarized Rape and Other Patriarchal Hostilities: Fueling and Legitimating Male Demand for a Sex Trade and also The Organization of Military Prostitution in Modern Times: Building a Sex Trade from Militarized Demand in Kathryn Farr (2004) *Sex Trafficking: The Global Market in Women and Children.* New York: Worth Publishers.

368 See Small Arms Survey (2007) The militarization of Sudan: a preliminary review of arms flows and holdings. Human Security Baseline Assessment. Sudan Issue Brief Number 6. April 2007. Retrieved June 27, 2007 from http://www.smallarmssurvey.org/index.html.

369 William D. Hartung and Bridget Moix (2000) Deadly Legacy: U.S. Arms to Africa and the Congo War. *Arms Trade Resource Center.* January 2000. Retrieved April 7, 2007 from http://www.worldpolicy.org/projects/arms/reports/congo.htm.

370 Brian Ross, David Scott, and Rhonda Schwartz (2005) U.N. Sex Crimes in Congo. *ABC News.* Retrieved July 17, 2007 from http://abcnews.go.com/2020/UnitedNations/story?id=489306&page=1.

371 Donna Hughes (2000) The Internet and Sex Industries: Partners in Global Sexual Exploitation. *Technology and Society Magazine* Spring 2000. Retrieved February 13, 2005 from http://www.uri.edu/artsci/wms/hughes/siii.htm. Hughes discussed reporting by John Markoff (1998) Indictment says mob is going high-tech: six men arrested in alleged Vegas computer scheme. *New York Times* October 17, 1998.

372 Thanks to Philip Williams, Professor of International Affairs, Public and Urban Affairs at the Ridgeway Center for International and Security Studies, University of Pittsburg for his assistance in formulating these ideas about prostitution/pornography/trafficking and terrorism. See Phil Williams (1999) (editor) *Illegal Immigration and Commercial Sex: The New Slave Trade.* Portland: Frank Cass.

373 See John Arquilla, David Ronfeldt and Michelle Zanini (1999) Networks, Netwar, and Information-Age Terrorism. In Ian O. Lesser, Ian Lesser, Bruce Hoffman, John Arquilla, David Ronfeldt, Michele Zanini (eds.) *Countering the New Terrorism.* Washington DC: Rand. Retrieved June 23, 2006 from http://64.233.167.104/search?q=cache:65sYO2SH_A0J:www.rand.org/pubs/monograph_reports/MR989/MR989.chap3.pdf+Networks+Netwar+and+Information+age+terrorism&hl=en&gl=us&ct=clnk&cd=.1.

374 For example, border crossing via bribes of immigration officials, use of air ambulances for trafficking women or drugs.

[375] Catharine A. MacKinnon (2006) Women's September 11th: Rethinking the International Law of Conflict. In *Are Women Human? and Other International Dialogues*. Cambridge: Harvard University Press, pages 259-278.

[376] Personal Communicatoin (2006) with Philip Williams, Professor of International Affairs, Public and Urban Affairs at the Ridgeway Center for International and Security Studies, University of Pittsburg. August 31, 2006.

Chapter 8. The Role of Cab Drivers in Las Vegas Prostitution

[377] Louise Brown (2000) *Sex Slaves: The Trafficking of Women in Asia*. London: Virago, page 13.

[378] The interview took place in March 2006.

[379] Lily Hyde and Marina Denisenko (1997) Modern-day slavery traps local women. *Kyiv Post* October 9, 1997. Cited in Donna M. Hughes (2000) The 'Natasha Trade:' The Transnational Shadow Market of Trafficking in Women. *Journal of International Affairs* 53(2): 625-651. Spring 2000.

[380] Jeff German (2005) Examining how strip clubs and drivers are working things out -- for now. *Las Vegas Sun* December 9, 2005. Retrieved March 27, 2006 from http://www.lasvegassun.com/sunbin/stories/commentary/2005/dec/09/519793694.html. See also Abigail Goldman (2007) Brothels and strip clubs get customers, cabbies get cash. *Las Vegas Sun* May 27, 2007. Retrieved May 29, 2007 from http://www.lasvegassun.com/sunbin/stories/lv-other/2007/may/27/566627574.html.

[381] Or should not drive. Personal communication (2006) with Howard Meadow, Las Vegas. March 19, 2006. Howard Meadow, who directs alternative sentencing programs for both substance abusers and johns at the Las Vegas Municipal Court, stated that Las Vegas has the highest fatality rate from DUIs (driving under the influence of alcohol or drugs) in the US. It also has the highest auto insurance rates. Meadow stated that in Las Vegas, 25% of all automobiles have a drunk driver at the steering wheel.

[382] George Knapp (2005) KLAS-TV.com Las Vegas July 1, 2005. *Cabbie Kickbacks Corruption* Part 2. Retrieved June 15, 2006 from http://www.klas-tv.com/Global/story.asp?S=3549362. See also B.K. Jordan (2004) *Stripped: Twenty Years of Secrets from Inside the Strip Club*. Kearney Nebraska: Satsu Multimedia Press, pages 123-129 where the cab driver lobby in Las Vegas is described from a strip club bouncer' s perspective.

[383] Knapp, *Cabbie Kickbacks Corruption* Part 2.

[384] Steve Miller (2007) Nevada Legalizes Extortion. American Mafia.com February 12, 2007. Retrieved July 16, 2007 from http://www.americanmafia.com/Inside_Vegas/2-12-07_Inside_Vegas.html.

385 Knapp, *Cabbie Kickbacks Corruption* Part 2. "They're pimping a ride so to speak."

386 Abilgail Goldman (2007) Brothels and strip clubs get customers, cabbies get cash. *Las Vegas Sun* May 27, 2007. Retrieved June 2, 2007 from http://www.lasvegassun.com/sunbin/stories/sun/2007/may/27/566627574.h tml?prostitution. See also Editorial (2007) Kickbacks from brothels: tourists discover taxi and limo drivers know more than simply where the best buffet is *Las Vegas Sun* May 30, 2007. Retrieved June 2, 2007 from http://www.lasvegassun.com/sunbin/stories/sun/2007/may/30/566676653.h tml?prostitution.

387 Adrienne Packer (2005) Trouble: Do clubs get taken for ride? Owners say cabbies force them to pay for patrons. *Las Vegas Review-Journal* October 29, 2005. Retrieved June 9, 2006 from http://www.reviewjournal.com/lvrj_home/2005/Oct-29-Sat-2005/news/4058257.htmlTIP. Assembly Bill 505 included language that would have prohibited cabdrivers from accepting tips from any person or business that had been issued a license from the county. See also Cy Ryan, Mary Manning, and Eric Leake (2005) Governor to Veto Cabbie Bill: Drivers threatened to strike over ban on business tips. *Las Vegas Sun* June 10, 2005. Retrieved May 22, 2006 from http://www.lasvegassun.com/sunbin/stories/text/2005/jun/10/518886820.ht ml.

388 Ed Vogel (2007) Guinn Veto Upheld: Assembly keeps cabbies' tips in place. *Las Vegas Review-Journal* February 9, 2007. Retrieved June 7, 2007 from http://www.reviewjournal.com/lvrj_home/2007/Feb-09-Fri-2007/news/12483899.html.

389 Personal Communication (2006) with Brent Jordan. Las Vegas, Nevada, March 23, 2006.

Chapter 9. Political and Judicial Corruption and the Nevada Sex Industry

390 Louise Brown (2000) *Sex Slaves: The Trafficking of Women in Asia*. London: Virago, page 186.

391 *The Hindu* Online Newspaper (2007) MPs' links to human trafficking serious matter: Somnath. New Delhi. April 21, 2007. Retrieved April 30, 2007 from http://www.hindu.com/thehindu/holnus/002200704211861.htm. See also Zeenews (2007) Katara aide names 3 more MPs. New Delhi, April 21 2007. Retrieved April 30, 2007 from http://www.zeenews.com/znnew/articles.asp?aid=367039&sid=NAT.

392 Person Communication (2006) with Nye County Commissioner Candice Trummell. July 7, 2006.

[393] CBS13 (2006) Nevada Politicians Accept Brothel Donations. July 18, 2006. Retrieved   July   9,   2007   from http://cbs13.com/local/local_story_199200239.html.

[394] Jeanie Kasindorf (1985) *The Nye County Brothel Wars*. New York: Dell. See pages 226-227.

[395] Joe Schoenmann (2003) Reviving Downtown on the backs of taxed and regulated prostitutes. *Las Vegas Weekly* July 10, 2003. Retrieved July 26, 2005 from http://www.lasvegasweekly.com/2003/07_10/news_coverstory.html.

[396] Department of Justice News Release (2006) Pahrump brothel owner arrested and charged with fraud for unlawful payments to Nye County Commissioner. March   6,   2006.   Retrieved   August   17,   2006   from http://lasvegas.fbi.gov/dojpressrel/pressrel06/wirefraud030606.htm.

[397] Personal Communication (2007) with Steve Miller. June 12, 2007.

[398] Kasindorf, *Nye County Brothel Wars*, page 300.

[399] Jordan, *Stripped*, page 94.

[400] Adrienne Packer (2006) Public Corruption Case: Galardi Describes '03 Fears. *Las Vegas Review-Journal* April 12, 2006. Retrieved July 27, 2006 from http://www.reviewjournal.com/lvrj_home/2006/Apr-12-Wed-2006/news/6827988.html.

[401] John. L. Smith (2006) It's hard keeping score when game goes from golf course to strip club. *Las Vegas Review-Journal* April 05, 2006. Retrieved July 27, 2006 from
http://www.reviewjournal.com/lvrj_home/2006/Apr-05-Wed-2006/news/6708540.html

[402] John L. Smith (2006) New Galardi allegations present a tale of two public officials. *Las Vegas Review-Journal* April 11, 2006. Retrieved July 27, 2006 from http://www.reviewjournal.com/lvrj_home/2006/Apr-11-Tue-2006/news/6804829.html.

[403] Case Number 03-470-JCM(LRL). The four Clark County commissioners were Mary Kincaid-Chauncey, Dario Herrera, Erin Kenny, and Lance Malone. See Adrienne Packer (2006) Kincaid-Chauncey, Herrera prepare to fight charges they took Galardi bribes. *Las Vegas Review-Journal* March 12, 2006. Retrieved   July   27,   2006   from http://www.reviewjournal.com/lvrj_home/2006/Mar-12-Sun-2006/news/6311788.html.

[404] A strip club bouncer complained that at "every city council meeting, the rules governing strip clubs change." For example, rules concerning whether or not customers can touch lap dancers or not, how far away they must be, how much of the dancer's breasts or buttocks can be shown. See Jordan, *Stripped*, page 118.

[405] Josh Longobardy (2006) In search of the truth about Erin Kenny, who let greed and ambition overcome her. *Las Vegas Weekly* Aug. 10, 2006. Retrieved November 22, 2006 from

http://www.lasvegasweekly.com/2006/08/10/erinkenny.html. See also John L. Smith (2006) Liz Renay's vice pales in comparison to today's sleaze. *Las Vegas Review-Journal* April 26, 2006. Retrieved July 27, 2006 from http://www.reviewjournal.com/lvrj_home/2006/Apr-26-Wed-2006/news/7051107.html.

[406] Donna M. Hughes (2000) The 'Natasha Trade:' The Transnational Shadow Market of Trafficking in Women. *Journal of International Affairs* 53(2): 625-651. Spring 2000. Citing data from Brunon Holyst (1999) Organized crime in Eastern Europe and its implications for the security of the Western World. In Stanley Einstein and Menachem Amir (eds.) *Organized Crime-Uncertainties and Dilemmas.* Chicago: The Office of International Criminal Justice, pages 67-93.

[407] Michael J. Goodman and William C. Rempel (2006) In Las Vegas, They're Playing With a Stacked Judicial Deck: Some judges routinely rule in cases involving friends, former clients and business associates -- and in favor of lawyers who fill their campaign coffers. *Los Angeles Times* June 8, 2006. Retrieved July 26, 2006 from http://www.latimes.com/news/nationworld/nation/la-na vegas8jun08,0,453414,full.story?coll=la-home-headlines.

[408] Sito Negron (2004) Solicitation case, license issue hang over Treasures. *Las Vegas Sun* August 31, 2004. Retrieved July 22, 2006 from http://www.lasvegassun.com/sunbin/stories/sun/2004/aug/31/517433042.ht ml?crazy%20horse%20too. A Las Vegas City Council member, Michael Mack was a "consultant" for Treasures strip club. See also Steve Miller's commentary on the extensive corruption connected with Las Vegas strip club Treasures, from its start to shutdown: Two Lousy Lawyers, a Conflicted Judge, and a Silent Mayor at AmericanMafia.com June 21, 2004. Retrieved October 30, 2005 from http://www.americanmafia.com/inside_vegas/6-21-04_Inside_Vegas.html.

[409] According to Steve Miller, *Two Lousy Lawyers, a Conflicted Judge, and a Silent Mayor*, Judge Assad accepted $15,000 in cash and in-kind contributions from contributors who have close ties to persons in the case. His two main contributors include a man who is under federal indictment for bribing public officials, and another who is facing federal indictment for racketeering.

[410] *The Hindu* (2007) MPs' links to human trafficking serious matter: Somnath New Delhi, April 22, 2007. Retrieved June 13, 2007 from http://www.hindu.com/thehindu/holnus/002200704211861.htm. See also Zee News Online (2007) Katara aide names 3 more MPs. April 22, 2007. Retrieved June 13, 2007 afrom http://www.zeenews.com/znnew/articles.asp?aid=367039&sid=NAT.

[411] Robert I. Friedman (1996) India's Shame: Sexual Slavery and Political Corruption. *The Nation* 262(14), page 8. April 8, 1996.

[412] Peter Landesman (2004) The Girls Next Door. *New York Times Magazine* January 25, 2004. Retrieved February 18, 2004 from http://www.nytimes.com/2004/01/25/magazine/25SEXTRAFFIC.html.

[413] Lucha Castro (2007) Panel on Judicial Reform. *Feminicide = Sanctioned Murder: Race, Gender and Violence in Global Context.* Conference sponsored by Chicana and Chicano Studies of The Center for Comparative Studies in Race and Ethnicity at Stanford University. Palo Alto California. April 9, 2007.

[414] Steve Miller (2007) Inside Vegas. *The Rick Rizzolo Connection.* Retrieved January 21, 2007 from http://www.stevemiller4lasvegas.com/RickRizzolo.html. In this article, Miller stated, "Rick Rizzolo made his fortune by paying off crooked cops, politicians, DAs and judges to keep a lid on his and his associate's criminal activities. It has now taken the FBI and IRS to clean up what we on a local level should have had the courage to stop years ago. On September 6, 2006, the Las Vegas City Council unanimously voted to permanently revoke Rizzolo's liquor license and fine him $2.2 million dollars. This could not have happened had Mayor Oscar Goodman presided over the hearing. To make sure he didn't, on July 10, I filed an Ethics Complaint against him pointing out his serious conflicts of interest regarding Rizzolo. Because Goodman was forced to abstain by my complaint, the playing field was leveled allowing Mayor Pro Tem Gary Reese to preside over the meeting and shut down the Crazy Horse. But that didn't last for long."

[415] J. Brunner (2007) Strip-club owners charged over campaign contributions. *Seattle Times* June 7, 2007. Retrieved June 9, 2007 from http://seattletimes.nwsource.com/html/localnews/2003737549_colacurcio07m.html.

[416] Brian Haynes (2005) Officer faces sex charges: Police say woman sought to avoid arrest *Las Vegas Review-Journal* October 5, 2005. Retrieved January 10, 2006 from http://www.reviewjournal.com/lvrj_home/2005/Oct-05-Wed-2005/news/3692777.html. Glenn Puit (2004) *Ex-Police Officer Sent To Prison For Probation Breach.* Las Vegas Review-Journal September 17, 2004. Retrieved June 20, 2006 from http://www.reviewjournal.com/lvrj_home/2004/Sep-17-Fri-2004/news/24791820.html. Glenn Puit, *Case against ex-detective dismissed,* Las Vegas Review-Journal, November 1 2005. Retrieved January 8, 2006 from http://www.reviewjournal.com/lvrj_home/2005/Nov-01-Tue-2005/news/4097055.html.

[417] J.M.Kalil (2003) Internal Affairs Probe: Sergeant punished over loan. Longtime friend of topless club owner accepted $15,000 to finance venture. *Las Vegas Review-Journal* August 6, 2003. Retrieved September 8, 2006 from http://www.reviewjournal.com/lvrj_home/2003/Aug-06-Wed-2003/news/21886517.html.

[418] Brown, *Sex Slaves,* page 208.

[419] Glen Puitt (2005) Federal Investigation: Club manager arrested - Indictment alleges racketeering at Crazy Horse Too. *Las Vegas Review-Journal* Thursday, January 20, 2005. Retrieved June 20, 2006 from http://www.reviewjournal.com/lvrj_home/2005/Jan-20-Thu-2005/news/25700796.html.

[420] United Nations Convention Against Torture and Other Cruel, Inhuman or Degrading Treatment or Punishment, Dec. 10, 1984, 1465 U.N.T.S. 85.

[421] Istanbul Protocol: Manual on the Effective Investigation and Documentation of Torture and Other Cruel, Inhuman or Degrading Treatment or Punishment (1999) August 9, 1999. The Istanbul Protocol states that "Sexual torture begins with forced nudity, which in many countries is a constant factor in torture situations. One is never so vulnerable as when one is naked and helpless. Nudity enhances the psychological terror of every aspect of torture, as there is always the background of potential abuse and rape or sodomy. Furthermore, verbal sexual threats, abuse and mocking are also part of sexual torture, as they enhance the humiliation and degrading aspects of it, all part and parcel of the procedure. Groping women is traumatic in all cases, and considered torture. There are some differences between sexual torture of men and sexual torture of women, but several issues apply to both." Retrieved September 1, 2004 from http://www.unhchr.ch/pdf/8istprot.pdf.

[422] Metin Basoglu, Maria Livanou, Cvetana Crnobaric (2007) Torture vs Other Cruel, Inhuman, and Degrading Treatment: Is the Distinction Real or Apparent? *Archives of General Psychiatry* 64:277-285.

[423] It was surprisingly difficult to obtain quotes on the prices for Las Vegas Yellow Pages advertising. Three people made at least 15 phone calls which were not returned, put off, or where prices were not quoted for one reason or another.

[424] See for example a defamation lawsuit brought against Steve Miller when he reported the facts about organized crime, political payoffs, and violence at the Crazy Horse Too strip club in Las Vegas. Steve Miller (2005) The Last Hours of Crazy Horse Too: On the Boulevard of Broken Dreams - -and Necks. April 18, 2005 Retrieved September 7, 2006 from http://www.americanmafia.com/Inside_Vegas/4-18-05_Inside_Vegas.html.

[425] Cynthia Cotts (2001) The Wrong Way to Get Screwed: Reporter Fired for Exposing Whorehouse. *Village Voice* April 18-24 2001. Retrieved May 6, 2007 from http://www.villagevoice.com/news/0116,cotts,23961,6.html.

Chapter 10. It's the Advertising, Dummy!

[426] Richard Abowitz (2001) Cathouse Dreams. *Las Vegas Weekly* June 7, 2001. Retrieved June 20, 2001 from http://72.14.253.104/search?q=cache:Fy0P7z5jRw4J:www.lasvegasweekly.com

/2001/features/06_07_CathouseDreams/cathouse_dreams.htm+Richard+Ab
owitz,+(2001)+Cathouse+Dreams&hl=en&ct=clnk&cd=1&gl=us.

[427] "… illegal prostitution does indeed flourish in Nevada. Commercial sex is offered by juveniles and adults, all of whom are marginalized and exploited in various ways and to various degrees as a result of their age, social class, race, ethnicity, and gender." Kathryn Hausbeck (2004) *Sex Industry and Sex Workers in Nevada.*   Retrieved   February   17,   2006   from http://www.unlv.edu/centers/cdclv/healthnv/sexindustry.html.

[428] Erin Neff (2003) Legalized Prostitution: Vegas brothels suggested - Goodman remarks open debate on downtown red-light district. *Las Vegas Review-Journal* October 24, 2003. Retrieved December 2, 2005 from http://www.reviewjournal.com/lvrj_home/2003/Oct-24-Fri-2003/news/22438503.html.

[429] July 2006 advertising on San Francisco radio station.

[430] http://www.visitlasvegas.com/vegas/features/be-anyone/index.jsp.

[431] Joel Stein (2004) The Strip is Back. *Time* July 26 2004.

[432] Mary Lucille Sullivan   (2007) *Making Sex Work: A Failed Experiment with Legalized Prostitution.* Spinifex: North Melbourne, page 8.

[433] Robin Kwong (2007) U.S. presses Macao over Human Trafficking. Retrieved July 4, 2007 from http://www.ft.com/cms/s/59af099e-27e5-11dc-80da-000b5df10621.html.

[434] K.C. Howard (2007) FBI says human trafficking not involved in brothel raids. *Las Vegas Review-Journal* May 10, 2007. Retrieved May 10, 2007 from http://www.lvrj.com/news/7436826.html.

[435] Brian Haynes (2007) Operation Doll House: Bust 'just tip of the iceberg': Authorities suspect women forced into prostitution. *Las Vegas Review-Journal* April   24,   2007.   Retrieved   April   28,   2007   from http://www.lvrj.com/news/7164646.html.

[436] Online reviews of Dutch escort agencies advertising for example "shy, embarrassed, and awkward young" Asian women and also the eastern European women mentioned in the text. Retrieved April 20, 2007 from http://www.ignatzmice.com/index.php?categoryid=42

[437] Lenore Skenazy (2007) Why do we let classified ads sell sex slaves? New York   Sun   July   7,   2007.   Retrieved   July   9,   2007   from http://www.nashuatelegraph.com/apps/pbcs.dll/article?AID=/20070707/OP INION04/207070325/-1/opinion.

[438] Table 6 was constructed in 2006-2007 by Kathy Watkins of Nitecat Media in San Francisco who also researched data contained in the table.

[439] Amy O'Neill Richard (2000) International Trafficking in Women to the United States: A Contemporary Manifestation of Slavery and Organized Crime. DCI Report, United States Department of State.

[440] Thank you Annie Fukushima of Berkeley, California for assistance in calculating this information from the Yellow Pages in Las Vegas.

441 Annie Fukushima, 2007, in a doctoral program in Women's Studies and Ethnic Studies at University of California, Berkeley, contributed this information from her research.

442 8th Street Latinas. http://www.8thstreetlatinas.com/main.htm?id=faxxaff.

443 Personal Communication (2007) with Roger Young, Reno Nevada. I spoke with Mr. Young about pornography on July 9, 2007. Young consults with various groups on the role of prostitution and pornography in promoting violence against women

444 *The Economist* (1998) The Sex Business. February 14, 1998, page 21.

445 Donna Hughes (2003) Prostitution Online. In Melissa Farley (ed.) *Prostitution, Trafficking and Traumatic Stress.* Binghamton: Haworth Press, pages 115-132.

446 Rebecca Whisnant (2007) Not Your Father's Playboy, Not Your Mother's Feminist Movement: Contemporary Feminism in a Porn Culture. Speech at Pornography and Pop Culture: Re-framing Theory, Re-thinking Activism. Boston MA, March 24, 2007.

447 See for example Akiyuki Nozaka (1970) *The Pornographers* (Michael Gallagher translation, 1968) Tokyo: Charles Tuttle Publishing.

448 Pornography has been used as a means of recruitment into childhood sexual assault as well as recruitment into prostitution and trafficking. See Catharine A. MacKinnon and Andrea Dworkin (1997) *In Harm's Way: The Pornography Civil Rights Hearings.* Cambridge: Harvard University Press. Pornography that normalizes prostitution is used by pimps to teach girls what acts to perform in prostitution. Mimi H. Silbert & Ayala M. Pines, Pornography and Sexual Abuse of Women (1987) *Sex Roles* 10: 857. Women in prostitution have described pornography's role in submitting to the enactment of specific scenes for pimps or customers. See also Melissa Farley & Howard Barkan (1998) Prostitution, Violence and Posttraumatic Stress Disorder. *Women & Health* 27: 37.

449 Robert Jensen (2006) The Paradox of Pornography. *Op Ed News.* Retrieved February 1, 2006 at http://www.opednews.com/articles/opedne_robert_j_060201_the_paradox_of_porno.htm (quoting Jeff Steward, owner of JM Productions, http://www.jerkoffzone.com).

450 Teela Sanders (2005) 'It's Just Acting:' Sex Workers' Strategies for Capitalizing on Sexuality. *Gender, Work and Organization* 12: pages 319, 330 .

451 Robin Askew (2000) Interview. *Sex: The Annabel Chong Story.* Retrieved March 16, 2006 from http://www.spikemagazine.com/1000annabelchong.php.

452 Askew, Sex: The Annabel Chong Story.

453 Francis Hanly (2005) *Debbie Does Dallas Uncovered: The True Story Behind the Most Famous Porn Film of All Time.* Sundance Channel Home Entertainment. See other information about Zaffarano's association with the Mitchell Brothers in John Hubner (1993) *Bottom Feeders.* New York: Doubleday.

454 This information about Jenna Jameson is well known. It is in her autobiography and on fan web sites. See http://www.popstarsplus.com/celebrities_jennajameson.htm

455 Melissa Farley (2007) Renting an Organ for Ten Minutes: What Tricks Tell us about Prostitution, Pornography, and Trafficking. In D. Guinn (ed.) *Pornography: Driving the Demand for International Sex Trafficking.* Los Angeles: Captive Daughters Press.

456 Catharine A. MacKinnon (2001) *Sex Equality.* New York: Foundation Press, pages 1520, 1524 .

457 Thomas Zambito (2006) Two Plead Guilty in $13 Million Prosty Ring. *New York Daily News.* Retrieved January 6, 2006 from http://www.nydailynews.com/front/story/380758p-323279c.html.

458 De A. Clarke. (2004) Prostitution for Everyone: Feminism, Globalization, and the 'Sex' Industry. In Rebecca Whisnant and Christine Stark (eds.) *Not for Sale: Feminists Resisting Prostitution and Pornography.* Melbourne: Spinifex.

459 Catharine A. MacKinnon (2006) Pornography as Trafficking. In *Are Women Human? and Other International Dialogues.* Cambridge: Harvard University Press, Cambridge: Harvard University Press, pages 247-258.

460 Kathryn Hausbeck and Barbara G. Brents (2000) Inside Nevada's Brothel Industry. In Ronald Weitzer (ed.) *Sex for Sale: Prostitution, Pornography and the Sex Industry.* New York: Routledge, page 230.

461 Brent K. Jordan (2004) *Stripped: Twenty Years of Secrets From Inside the Strip Club.* Kearny Nebraska: Satsyu Multimedia Press, page 107.

462 http://tour.stripclubnetwork.com/.

463 Retrieved June 23, 2007 from http://lasvegas.craigslist.org/w4m/. On June 23, 2007 for example, an ad for prostitution at http://www.geocities.com/lonelywifelulu/ was placed on craigslist Las Vegas under the category "woman seeking man."

464 Internet prostitution and pornography offer the john anonymity. There are increasing numbers of online communities of johns or as *Playboy* July 2007 describes them - 'hobbyists' - supporting each others' predatory behaviors and exchanging information regarding where and how women can be bought. See Sven Axel Mansson (2004) Men's Practices in Prostitution and Their Implications for Social Work. In Sven Axel Mansson and Clotilde Proveyer (eds.) *Social Work in Cuba and Sweden.* University of Havana.

465 Peter Landesman (2004) The Girls Next Door. *New York Times Magazine* January 25, 2004. Retrieved February 18, 2004 from http://www.nytimes.com/2004/01/25/magazine/25SEXTRAFFIC.html.

466 One such site is http://www.cams.com. The parent company of cams.com is Steamray (http://www.streamray.com). In January 2006 Steamray announced the broadcast of 400 pornographic videos simultaneously with the goal of becoming the world's largest webcam company. (http://www.avnonline.com/index.php?Primary_Navigation=Editorial&Actio

n=View_Article&Content_ID=255014). Steamray subsequently merged with Various, the parent company of Adultfriendfinder.com, owned by Andrew Conru. Adultfriendfinder features gonzo pornography. See note 460. In August 2005, Adultfriendfinder claimed that 17 million people visit the site, and that the company employs 200 people. See http://www.avnonline.com/index.php?Primary_Navigation=Editorial&Action =View_Article&Content_ID=235604. Other sites owned by Various are AdultFriendFinder.com, FriendFinder.com, Alt.com, OutPersonals.com, Passion.com, GradFinder.com, NiceCards.com, QuizHappy.com, BreakThru.com, Dine.com, BigChurch.com, ShareRent.com, FriendPages.com, FilipinoFriendFinder.com AsiaFriendFinder.com, GUANXI.com, IndianFriendFinder.com, SeniorFriendFinder.com JewishFriendFinder.com, Amigos.com, GermanFriendFinder.com, FrenchFriendFinder.com, KoreanFriendFinder.com

[467] Confidential law enforcement source, 2005.

[468] Gonzo pornography is extremely violent pornography where women are violently raped, obviously injured, painfully tied up, often terrified and/or crying. See Martin Amis (2001) A Rough Trade: Martin Amis reports from the high-risk, increasingly violent world of the pornography industry. *UK Guardian*. Saturday March 17, 2001. Retrieved April 9, 2005 from http://books.guardian.co.uk/departments/politicsphilosophyandsociety/story/ 0,,458058,00.html.

[469] Stuart Millar (2000) Sex gangs sell prostitutes over the internet. *UK Guardian* July 16, 2000.

[470] For example pimp/pornographer Dennis Hof in a brothel near Carson City, Nevada which was promoted by Larry Flynt, Bob Guccione, and Howard Stern.. Hof hired women in pornography to prostitute at his brothel and advertised them for sale at higher prices. On the other hand Hof put intense pressure on prostituting women to be filmed even when they did not want that. One woman told me that unless the women at Hof's brothel agreed to be filmed they were unable to earn the alleged income advertised as easily obtainable at the brothel. Since many women enter prostitution as a last-ditch survival effort, they resist being filmed because that would be a record of their prostitution, and often a part of their lives that they prefer to leave behind them, rather than have on view indefinitely into the future.

[471] Richard Abowitz (2005) Aurora Snow previews the AVN Awards. Dispatches from Las Vegas. *Los Angeles Times* December 30, 2005. Retrieved January 7, 2006 from http://vegasblog.latimes.com/vegas/2005/12/index.html

[472] Adult Video News (2006) Porn in Las Vegas, quoting an article from *Las Vegas Weekly* by Richard Abowitz. Retrieved January 7, 2006 from http://www.avn.com/index.php?Primary_Navigation=Articles&Action=Print _Article&Content_ID=21324.

473 Ray Pistol has been in the business of filming sexual exploitation for many years. His company produces pornography based on themes in the original pornography of the rapes of Linda Marchiano in Deep Throat.

474http://www.searchextreme.com/directors/Thomas_Zupko/35630306841/.

475 Richard Abowitz (2000) Refugees from L.A.'s adult film industry are making Las Vegas home. *Las Vegas Weekly* May 11, 2000. Retrieved June 17, 2005 from http://lasvegasweekly.com/features/up_and_coming.html. Producer Christi Lake shot Fan Fuxxx in Las Vegas in which actors "included a cocktail waitress, a pilot and a pit boss from the Mirage." Lake appreciated Nevada's permissive law and cheap permits to shoot pornography: "if I'm shooting on private property, I don't have to go and get a permit, which is wonderful."   Adult Video News (2006) Porn in Las Vegas, quoting an article from *Las Vegas Weekly* by   Richard   Abowitz   located   at http://www.avn.com/index.php?Primary_Navigation=Articles&Action=Print _Article&Content_ID=21324.

Chapter 12. Barriers to Service Provision for Women Escaping Nevada Prostitution and Trafficking

476 Marisa Ugarte, Laura Zarate, and Melissa Farley (2003) Prostitution and Trafficking of Women and Children from Mexico to the United States. In Melissa Farley (ed.) *Prostitution, Trafficking, and Traumatic Stress*. Binghamton: Haworth Press, pages 147-166. See this article for case examples of service provision to trafficked women and children.

Chapter 13. Attitudes toward Prostitution & Sexually Coercive Behaviors of Men at University of Nevada, Reno

477 American Psychological Association (2007) *Report of the APA Task Force on the Sexualization of Girls*. Retrieved March 20, 2007 from http://www.apa.org/pi/wpo/sexualization.html, page 28.

478 L.M. Ward (2003) Understanding the Role of Entertainment Media in the Sexual Socialization of American Youth: A Review of Empirical Research. *Developmental Review* 23: 347-388.   Neil M. Malamuth, Tamara Addison, and Mary Koss (2000) Pornography and Sexual Aggression: Are There Reliable Effects and Can We Understand Them? *Annual Review of Sex Research* 11: 26-91.

479 Ann Cotton, Melissa Farley, & Robert Baron (2002) Attitudes toward prostitution and acceptance of rape myths. *Journal of Applied Social Psychology* 32: pages 1-8.

480 Mary Koss (1988) Hidden rape: Sexual aggression and victimization in a national sample of students in higher education. In A.W. Burgess (ed.) Rape and Sexual Assault II. New York: Garland, pages 3-25; K. A. Lonsway and L.F.

Fitzgerald (1994) Rape Myths: in review. *Psychology of Women Quarterly* 18: 133-164; N.M. Malamuth, R.J. Sockloskie, M.P. Koss, and J.S. Tanaka (1991) Characteristics of aggressors against women: Testing a model using a national sample of college students. *Journal of Consulting and Clinical Psychology* 59: 670-681; J. Miller and M.D. Schwartz (1995) Rape myths and violence against street prostitutes. *Deviant Behavior: An Interdisciplinary Journal* 16: 1-23.

[481] For a description of rape myths see Women Against Violence Against Women website http://www.wavaw.ca/informed_myths.php.

[482] Patricia Tjaden and Nancy Thonnes (2000) Prevalence and Consequences of Male-to Female and Female-to-Male Intimate Partner Violence as Measured by the National Violence Against Women Survey. *Violence Against Women* 6 (2): 142-161

[483] B.L. Vogel (2000) Correlates of pre-college males' sexual aggression: Attitudes, beliefs and behavior. *Women and Criminal Justice* 11: 25-47; D.M. Truman, D.M. Tokar, and A.R. Fischer (1996) Dimensions of Masculinity: Relations to Date Rape, Supportive Attitudes, and Sexual Aggression in Dating Situations. *Journal of Counseling and Development* 74: 555-562; J.T. Spence, M. Losoff, and A.S, Robbins (1991) Sexually Aggressive Tactics in Dating Relationships: Personality and Attitudinal Correlates. *Journal of Social and Clinical Psychology* 10: 289-304.

[484] Cotton, Farley, and Baron (2002). r=.27, p<.0001.

[485] Megan Schmidt, Ann Cotton, and Melissa Farley (2000) Attitudes toward prostitution and self-reported sexual violence. Presentation at the 16th Annual Meeting of the International Society for Traumatic Stress Studies, San Antonio, Texas, November 18, 2000.

[486] Ann Cotton, Melissa Farley, and Megan Schmidt (2001) Prostitution Myth Acceptance, Sexual Violence, and Pornography Use. Presentation at Annual Meeting of the American Psychological Association, San Francisco CA. August 27, 2001.

[487] University of Nevada, Reno Office of Institutional Research.

[488] Reno, Nevada is located in Washoe County. Reno is 15 miles from the nearest legal brothel which is located in nearby Storey County. By state law, legal prostitution is not permitted in any county with a population greater than 400,000.

[489] Erin Meehan Breen (2004) Conforte in Exile. *Reno Gazette-Journal* May 29, 2004. Retrieved June 25, 2005 from http://www.rgj.com/news/stories/html/2004/05/29/71913.php.

[490] *Reno Gazette-Journal* (2000) Mustang Ranch Chronology. Aug. 8, 2000. Retrieved October 7, 2005 from http://www.rgj.com/news/stories/html/2002/11/29/29072.php.

[491] Because we ran post-hoc tests, we used the Bonferroni adjustment for multiple comparisons.

[492] We found significant differences between the attitudes of the Reno students toward prostitution and those of the other college students. $F(46, 231) = 3.37$, p=.000; Wilks' Lambda = .60; partial eta squared = .40.

[493] "There is nothing wrong with prostitution." $(F = 10.38, df 1, 276, p = .001)$.

[494] "Prostitution should be treated no differently than any other business." $(F=18.51, df 1, 276, p=.000)$.

[495] "Prostitution should be legalized." $(F = 15.47, df 1, 276 \ \ p = .000)$ or decriminalized $(F = 8.85, df = 1, 276, p = .003)$.

[496] "There is nothing wrong with having sex for money." $(F = 10.94, df = 1,276, p = .001)$.

[497] "Arresting men who patronize prostitutes causes more problems than it solves." $(F = 10.37, df=1,276, p = .001)$.

[498] "It's OK for a man to go to a call girl if his wife doesn't find out." $(F=6.44, df=1,383, p=.01)$.

[499] "I would use an escort service or patronize a call girl if I knew it was safe to do so." $F=4.47, df 1,379, p=.03)$.

[500] $F=5.79, df 1, 367, p = .02$.

[501] $F = 6.00, df = 1,377, p = .02$.

[502] Ron Levant (1997). Nonrelational sexuality in men. In Ron Levant and Gary Brooks (eds) *Men and Sex: New Psychological Perspectives*. New York: John Wiley, pages 9-27.

[503] "Prostitution lowers the moral standards of a community." $F=33.37, df 1,276, p = .000$.

[504] Melissa Farley (2007) 'Renting an Organ for 10 Minutes:' What Tricks Tell us about Prostitution, Pornography, and Trafficking. In D. Guinn (ed.) *Pornography: Driving the Demand for International Sex Trafficking*. Los Angeles: Captive Daughers Media.

[505] Prostitution sex makes men better lovers. $F=9.03, df 1,276, p = .003$. In fact, women in prostitution 'teach' men to ejaculate as quickly as possible because they dislike the sex of prostitution. Like other women, women in prostitution prefer having sex with men they love, not with anonymous strangers. Men are 'taught' in prostitution that women truly love a premature ejaculator, while at home their wives and girlfriends are left sexually frustrated by the dysfunction.

[506] $F=9.03, df 1,276, p = .003$

[507] "Prostitution is a choice that women should have." $F=9.03, df 1,276, p = .003$

[508] "It would be acceptable if my son went to prostitutes." $(F = 11.98, df= 1, 276, p = .001)$.

[509] "It would be acceptable if my son went to brothels." $(F=33.37, df=1,276, p=.000)$.

[510] "It would be acceptable if my daughter grew up to be a prostitute." $(F = 3.36, df = 1, 276, p = .012)$.

[511] "It is ridiculous for a call girl to claim she's been raped by a customer." (F = 12.88, df1,276, p = .000).

[512] "If a man pays for sex, the woman should do whatever he wants." (F = 5.57, df 1,276, p = .019).

[513] The Reno men significantly *less often* endorsed the statement that prostitution is an exploitation of women's sexuality. (F=4.08, df 1,276, p = -.044).

[514] Reno men were more likely to endorse rape myths. F(19, 341) = 2.27, p=.002; Wilks' Lambda = .88; partial eta squared = .11.

[515] "Women generally find being physically forced into sex a real 'turn-on.'" (F=5.12, df 1,359, p=.024).

[516] If a woman is willing to make out with a guy then it's no big deal if he goes a little further and has sex," (F=17.42, df 1, 359, p = .000).

[517] "Men from nice middle class homes almost never rape." (F=3.92, df 1,359, p= .014).

[518] Reno men more often used women in prostitution. (F = 8.00, df 1,12, p=.015).

[519] Reno men more often went to strip clubs. F= 4.25, df 1, 384, p = .040

[520] Reno men more often went to massage parlor brothels. F=5.56, df 1, 384, p = .019.

[521] Reno men watched more video pornography. F = 39.52, df 1,383, p=.000.

[522] Reno men watched more Internet pornography. F=18.32, df 1, 385, p = .000.

[523] *FBI Uniform Crime Report* (2004) Table 6. http://www.fbi.gov/ucr/cius_04/.

[524] Women in prostitution are frequently raped. They are often raped by johns who assume that the payment of money entitles men to the right to rape them, which is exactly what the Reno college students told us that they believed. See Melissa Farley, Ann Cotton, Jacquelyn Lynne, Sybile Zumbeck, Frida Spiwak, Maria E. Reyes, Dinorah Alvarez, and Ufuk Sezgin (2003) Prostitution and Trafficking in 9 Countries: Update on Violence and Posttraumatic Stress Disorder. *Journal of Trauma Practice* 2 (3/4): 33-74. Also in Melissa Farley (ed.) *Prostitution, Trafficking and Traumatic Stress* (2003) Binghamton: Haworth Press for a summary of studies documenting the extremely high incidence of rapes of women in prostitution. Available at http://www.prostitutionresearch.com/c-prostitution-research.html

14. Adverse Effects of a Prostitution Culture on Nonprostituting Women

[525] Claude Jaget (1980) *Prostitutes – Our Life*. Bristol: Falling Wall Press, page 21.

[526] Knapp, George (2006) Prostitution Crackdown in Las Vegas is 'Just The Beginning' KLAS-TV Report. Retrieved June 30, 2006 from http://www.klas-tv.com/Global/story.asp?S=5101862.

[527] Puengs Vongs (2005) Leers and Loathing in Las Vegas - Why Am I Mistaken for an Asian Sex Import? *New America Media.* Retrieved December 7, 2005 http://news.pacificnews.org/news/view_article.html article_id=1c00b44ed4efa4c23fe8fa6ce4bcccf4.

[528] Account of a young woman to whom this occurred. Confidential Interview, 2005.

[529] A prostitution researcher in Las Vegas commented that she didn't plan to study prostitution when she moved to Las Vegas seven years ago. But after she was propositioned three times in nine months, that changed. "I decided I had to either study it," she said, "or move." Josh Bohling (2004) Sex and the State. *The Shorthorn Online,* University of Texas at Arlington. Retrieved July 8, 2006 from http://www.theshorthorn.com/archive/2004/spring/04-apr-01/n040104-05.html.

[530] American Psychological Association (2007) *Report of the APA Task Force on the Sexualization of Girls.* Retrieved March 20 2007 from http://www.apa.org/pi/wpo/sexualization.html.

[531] American Psychological Association, *Task Force on the Sexualization of Girls,* pages 14-16.

[532] Claire Hoffman (2006) Joe Francis' Baby Give Me A Kiss. *Los Angeles Times* August 6, 2006. Retrieved August 7, 2006 from http://www.latimes.com/features/magazine/west/la-tm-gonewild32aug06,0,2664370.story?coll=la-home-headlines.

[533] Edward G. Armstrong (2001) Gangsta Misogyny: a content analysis of the portrayals of violence against women in rap music 1987-1993. *Journal of Criminal Justice and Popular Culture.* 8(2): 96-126.

[534] Kimberle Crenshaw (1993) Beyond Racism and Misogyny: Black Feminism and 2 Live Crew. In Mari J. Matsuda, Charles R. Lawrence, and Richard Delgado (eds.) *Words that Wound.* Boulder, CO: Westview Press, pages 111-132. A remarkable DVD was released in 2006, critiquing the misogyny and homophobia in hip hop: Directed by Byron Hurt, Hip Hop: Beyond Beats and Rhymes.

[535] http://www.flakmag.com/web/lilpimp.html.

[536] John L. Smith (2001) Las Vegas has a long way to go in dealing with sexual harassment. *Las Vegas Review-Journal* Wednesday April 25, 2001. Retrieved May 22, 2005 from http://www.reviewjournal.com/lvrj_home/2001/Apr-25-Wed-2001/news/15937128.html.

[537] Smith, Las Vegas has a long way to go in dealing with sexual harassment.

[538] Ann C. McGinley (2006) Harassment of Sex(y) Workers: Applying Title VII to Sexualized Industries. *Yale Journal of Law & Feminism* 18 (1): 65-108, page 70. See note #17 where McGinley discusses Kelly Ann Cahill's support for the Hooters' extremely hostile workplace for women.

[539] McGinley, Harassment of Sex(y) Workers, Note #76.

[540] McGinley, *Harassment of Sex(y) Workers,* page 88, quoting Rod Smith.

541 John L. Smith (2000) 'Stationize' or else: Many longtime Santa Fe cocktail waitresses await fate. *Las Vegas Review-Journal* September 20, 2000. Retrieved October 8, 2005 from http://www.reviewjournal.com/lvrj_home/2000/Sep-20-Wed-2000/news/14417075.html.

542 See Casinowatch's website for a collection of articles referring to these and similar examples. http://www.casinowatch.org/sex_prostitution/sexessful_gambling.html. For example, a waitress won the right to a trial against Dennis Rodman who grabbed her breasts when he served him a drink.

543 Joel Stein (2004) The Strip is Back. *Time* July 26, 2004.

544 In addition to offering big spenders ('whales') free hotel rooms and free drinks, casino hosts 'comp' high rollers with women in prostitution.

545 McGinley, Harassment of Sex(y) Workers, page 75, citing Jin Kim, Can an Employer Create a Hostile Working Environment Through Its Advertisements? Unpublished manuscript.

546 McKinley, *Harassment of Sex(y) Workers*, page 76, citing Jin Kim.

547 Personal Communication (2006) with Kelly Langdon, Nevada State Rape Prevention Coordinator. July 31, 2006.

548 *FBI Uniform Crime Report* (2004) Table 6 Retrieved July 22, 2006 from http://www.fbi.gov/ucr/cius_04/.

549 Mary Lucille Sullivan (2007) *Making Sex Work: A Failed Experiment with Legalized Prostitution*. Spinifex: North Melbourne, page 179.

550 See bulbul.com.

Chapter 15. Conclusion: Legalization of Prostitution, a Failed Experiment

551 Sam Skolnik (2007) Teen Prostitution Scourge Grows: Special court tries to help youngsters, not punish them. *Las Vegas Sun* January 7, 2007. Retrieved February 20, 2007 from http://lasvegassun.com/sunbin/stories/sun/2007/jan/07/566687059.html.

552 Juliet W. Casey (2005) Human traffic targeted: authorities launch second campaign. *Las Vegas Review-Journal* Wednesday March 16, 2005. Retrieved September 9, 2005 from http://www.reviewjournal.com/lvrj_home/2005/Mar-16-Wed-2005/news/26081825.html.

553 Kathryn Farr (2005) *Sex Trafficking: The Global Market in Women and Children.* New York: Worth Publishers, pages 56-57.

554 Bob Herbert (2006) Why Aren't We Shocked? *New York Times* October 16, 2006. Retrieved February 2, 2007 from http://select.nytimes.com/2006/10/16/opinion/16herbert.html?hp.

555 Herbert, 2006.

556 Abigail Goldman (2007) Wealthy men, willing women. *Las Vegas Sun* June 12, 2007. Retrieved June 13, 2007 from http://www.lasvegassun.com/sunbin/stories/sun/2007/jun/12/566634938.html?seekingarrangement.com.

557 American Psychological Association (2007) *Report of the APA Task Force on the Sexualization of Girls*, page 34. Retrieved March 20, 2007 from http://www.apa.org/pi/wpo/sexualization.html.

558 Kurt Eichenwald (2005) Through his Webcam, a boy joins a sordid online world. *New York Times* December 19, 2005. Retrieved December 29, 2005 from http://www.nytimes.com/2005/12/19/national/19kids.ready.html?ei=5070&en=6915868c2227c0f9&ex=1181707200&adxnnl=1&adxnnlx=1181587446-EQqaCtbzIeK55+2a3EqssA.

559 Andrew Vacchs (2005) Watch Your Words. *Parade*. Retrieved April 30, 2007 from http://www.vachss.com/av_dispatches/parade_060505.html.

560 Cathy Spitz Widom (1995) Victims of Childhood Sexual Abuse – Later Criminal Connections. *National Institute of Justice Research in Brief.* March 1995. Retrieved July 8, 2007 from http://72.14.253.104/search?q=cache:8RnjY_C-9lgJ:www.ncjrs.gov/pdffiles/abuse.pdf+site:www.ncjrs.gov+Victims+of+Childhood+Sexual+Abuse+%E2%80%93+Later+Criminal+Connections.+National+Institute+of+Justice+Research+in+Brief.&hl=en&ct=clnk&cd=1&gl=us. Thank you Sharon Cooper M.D., FAAP for reminding me of the importance of Widom's research on the connections between childhood sexual assault and later prostitution.

561 Kim Curtis (2007) Sex Offenders Younger, More Violent. *Baltimore Sun* June 9, 2007. Retrieved June 12, 2007 from http://www.baltimoresun.com/news/nationworld/nation/wire/sns-ap-youth-sex-offenders,0,4681135.story.

562 Ryan Kim (2005) Bump, grind your way to riches, students told. *San Francisco Chronicle*. January 14, 2005. Retrieved January 14, 2005 from http://www.fradical.com/Pimping_at_school_career_day.htm.

563 Brian Haynes (2007) Cheerleading Coach Boasted of Girls, Clients, Report Says. *Las Vegas Review Journal.* June 12, 2007. Retrieved June 29, 2007 from http://www.lvrj.com/news/7953447.html?numComments=1000.

564 Louise Brown (2000) *Sex Slaves:The Trafficking of Women in Asia*. London: Virago, page 60.

565 Diane Matte of the Quebec organization Concertation des luttes contre l'exploitation sexuelle (Coalition against Sexual Exploitation) provided this information to me on September 22, 2005 in Montreal.

566 Vednita Carter and Evelina Giobbe (1999) Duet: Prostitution, Racism, and Feminist Discourse. *Hastings Women's Law Journal* 10 (1):37-58. Winter 1999.

567 Lenore Kuo (2002) Prostitution Policy: Revolutionizing Practice through a Gendered Perspective. New York University Press, pages 132-133.

568 Pimp Joe Richards was charged with wire fraud and bribery of Trummell in 2006. Gina B. Good (2006) Bribery, wire fraud charges: Joe Richards stung by FBI. *Pahrump Valley Times* March 8, 2006. Retrieved March 18, 2006 from http://www.pahrumpvalleytimes.com/2006/03/08/news/richards.html. See also Doug McMurdo (2006) Richards pleads not guilty in federal court. *Pahrump Valley Times* April 12, 2006. Retrieved April 13, 2007 from http://www.pahrumpvalleytimes.com/2006/04/12/news/richards.html. See also Department of Justice (2006) Pahrump Brothel Owner Indicted on Fraud Charges. Retrieved May 17, 2006 from http://www.usdoj.gov/usao/nv/home/pressrelease/march2006/richards0322 06.htm.

569 Marjan Wijers and Lin Lap-Chew (1997) *Trafficking in Women, Forced Labour and Slavery-like Practices in Marriage, Domestic Labour and Prostitution.* Utrecht: Foundation against Trafficking in Women, page 152.

570 Wijers and Lap-Chew, *Trafficking in Women*, page 199. According to reports from over 100 non-governmental organizations, these abuses are perpetrated four times more often against illegal migrant women than among national and legal migrant women in Europe.

571 Suzanne Daley (2001) New Rights for Dutch Prostitutes, but No Gain. *New York Times* August 12, 2001. Retrieved August 25, 2001 from http://www.nytimes.com/2001/08/12/international/12DUTC.html.

572 Janice G. Raymond (2004) Ten Reasons for *Not* Legalizing Prostitution And a Legal Response to the Demand for Prostitution. In Melissa Farley (ed.) *Prostitution, Trafficking, and Traumatic Stress.* Binghamton: Haworth Press, page 317. Also available at http://action.web.ca/home/catw/readingroom.shtml?x=81465.

573 Gerben J.N. Bruinsma and Guus Meershoek (1999) Organized Crime and Trafficking in Women from Eastern Europe in the Netherlands. In Phil Williams (ed.) *Illegal Immigration and Commercial Sex: The New Slave Trade.* Portland: Frank Cass, page 107. See also Raymond, *Ten Reasons for Not Legalizing Prostitution.*

574 Nye County Nevada May Outlaw Prostitution (2005) Retrieved March 12, 2007 from http://www.swop-usa.org/news/NyeCounty.php.

575 Author's italics. Josh Johnson (2007) Legal Prostitution is an Antiquated Joke. *Lahontan Valley News and Eagle Standard.* April 7, 2007. Retrieved July 27, 2007 from http://www.lahontanvalleynews.com/article/20070407/OPINION/10407003 9&SearchID=73288452865482. Johnson is editor of the Latontan Valley News and Eagle Standard I Fallon, located in Churchill County, Nevada. Brothels are legal in Churchill County but none have been in operation since 2004.

576 Nye County Nevada May Outlaw Prostitution (2005) Retrieved March 12, 2007 from http://www.swop-usa.org/news/NyeCounty.php.

[577] Adam Tanner (2006) Nevada gives legalized prostitution uneasy embrace. Retrieved Monday Feb 13, 2006 from http://today.reuters.co.uk/news/newsArticle.aspx?type=reutersEdge&storyID =2006-02-13T144614Z_01_ZWE353157_RTRUKOC_0_LIFE-PROSTITUTION.xml.

[578] Ray Hagar of the Reno Gazette-Journal asked Hillary Clinton, "Republican presidential candidate Mitt Romney has called Nevada's system of legalized prostitution "repugnant." You are a big advocate for women's rights so would you offer an opinion on Nevada's system of legalized prostitution?" Clinton replied: "I do not approve of legalized prostitution or any kind of prostitution. It is something that I personally believe is demeaning to women... I would obviously speak out against prostitution and try to persuade women that it is not - even in a regulated system - necessarily a good way to try to make a living. Let's try to find other jobs that can be there for women who are looking for a good way to support themselves and their families." Ray Hagar (2007) Conversation with Sen. Clinton. *Reno Gazette-Journal* April 30 2007. Retrieved May 3, 2007 from http://news.rgj.com/apps/pbcs.dll/article?AID=/20070429/NEWS10/70429 0370/1016/NEWS.

[579] Letter sent by Steve Wynn to all Nevada state senators and assembly members in 1988. Quoted by Alexa Albert in *Brothel* (2001) New York: Random House, page 180.

[580] Richard R. Becker and Ellen Levine (1994) Taking the Sin out of Sin City: a look at prostitution in Nevada. *Gauntlet* 1: 33-40, page 36.

[581] Erin Neff (2003) Legalized Prostitution: Vegas brothels suggested - Goodman remarks open debate on downtown red-light district. *Las Vegas Review-Journal* October 24, 2003. Retrieved July 22, 2006 from http://www.reviewjournal.com/lvrj_home/2003/Oct-24-Fri-2003/news/22438503.html. See also," I'm going to urge all able-bodied constituents to go out and have a lap dance," Goodman says, grinning. (John L. Smith (2001) Downturn depresses tease-and-tassel, but skin joints seek no stimulus. *Las Vegas Review-Journal* Sunday November 18, 2001. Retrieved September 19, 2006 from http://www.reviewjournal.com/lvrj_home/2001/Nov-18-Sun-2001/news/17469145.html.

[582] Joel Stein (2004) The Strip is Back. *Time.* July 26, 2004.

[583] Ed Koch (2003) Mayor reiterates brothels would help downtown. *Las Vegas Sun* October 24, 2003. Retrieved June 13, 2007 from http://www.lasvegassun.com/sunbin/stories/lv-gov/2003/oct/24/515777670.html.

[584] Lois R. Helmbold (2005) Women's Studies in Sin City: Reactionary Politics and Feminist Possibilities. *National Women's Studies Association Journal.* Summer 2005.

[585] See extensive documentation in John L. Smith (2005) *Sharks in the Desert: The Founding Fathers and Current Kings of Las Vegas*. Ft Lee, NJ: Barricade. See also John L Smith (2003) *Of Rats and Men: Oscar Goodman's Life from Mob Mouthpiece to Mayor of Las Vegas*. Las Vegas: Huntington Press.

[586] Neff, Legalized Prostitution.

[587] Sullivan, *Making Sex Work*, page 135.

[588] Hakim Almasmari (2005) Poverty + Tourism = Prostitution. *Yemen Times* 14 (892) Nov 7-9, 2005. Retrieved January 8, 2006 from http://yementimes.com/article.shtml?i=892&p=report&a=2.

[589] Mary Sullivan (2007) Whose Rights are We Talking About? Online Opinion. *Arena*. Retrieved June 25, 2007 from http://www.onlineopinion.com.au/view.asp?article=6011

[590] Personal Communication (2006) with Howard Meadow. Las Vegas, March 19, 2006.

[591] Sullivan, *Making Sex Work*, pages 163, 241.

[592] Alesia Adams (2004) Speech at National Conference on Domestic Trafficking and Prostitution, U.S. Department of Justice, Tampa, Florida, July 15-17, 2004. Adams was then Director of The Center to End Adolescent Sexual Exploitation in Atlanta (CEASE). She is now Territorial Services Coordinator Against Human and Sexual Trafficking at the Salvation Army in Atlanta.

[593] Girls' Educational and Mentoring Services in New York. See http://www.gems-girls.org/.

[594] Clyde Haberman (2007) The Young and Exploited Ask for Help. *New York Times* June 12, 2007. Retrieved June 15, 2007 from http://select.nytimes.com/search/restricted/article?res=F00A10FA3C5B0C71 8DDDAF0894DF404482.

[595] For example WestCare in Las Vegas operates programs for runaway youths and offers training for other agencies likely to encounter these children.

[596]See description of peer-led groups in two chapters in Melissa Farley (ed.) *Prostitution, Trafficking and Traumatic Stress* (2003): Jannit Rabinovitch "PEERS: The Prostitutes' Empowerment, Education and Resource Society," pages 239-254. And Norma Hotaling, Autumn Burris, B. Julie Johnson, Yoshi M. Bird, Kirsten A. Melbye "Been There Done That: SAGE A Peer Leadership Model among Prostitution Survivors," pages 255-266.

[597] Personal Communication (2005) with Kathleen Boutin at Independent Living Center, Nevada Partnership for Homeless Youth.

[598] Jack Sheehan (2004) *Skin City: Uncovering the Las Vegas Sex Industry*. Las Vegas: Stephens Press, page 184.

[599] Amanda, Roxburgh, Louisa Degenhardt, and Jan Copeland (2006) Posttraumatic stress disorder among female street-based sex workers in the greater Sydney area, Australia. *BMC Psychiatry* Volume 6. Open Access Online Journal at

http://www.pubmedcentral.nih.gov/articlerender.fcgi?artid=1481550. In New South Wales, Australia, where Sydney is located, 75% of the women prostituted legally in the street, 67% in cars, and only 57% chose to prostitute in what are called safe houses, but seem to actually be legal brothels. Thus, these women in NSW preferred legal street prostitution over legal brothel prostitution.

600 Roxburgh, Degenhardt, and Copeland, Posttraumatic stress disorder among female street-based sex workers in the greater Sydney area.

601 Brown, *Sex Slaves*, page 188.

602 Kate Hausbeck (2001) quoted in Richard Abowitz (2001) Cathouse Dreams *Las Vegas Weekly* June 7, 2001. Author's italics. Retrieved September 19, 2005 from http://www.lasvegasweekly.com/2001/features/06_07_CathouseDreams/cath ouse_dreams.htm.

603 Jaget, *Prostitutes- Our Life*, pages 102-103.

604 Sullivan, *Making Sex Work*, page 8.

605 The Swedish law on prostitution is now included in the Swedish Criminal Code, Chapter 6, Sexual Crimes section 11.

606 Women should be permitted to sue johns and pimps for damages caused by their acts. Catharine A. MacKinnon has described this approach in a chapter, "Prostitution and Civil Rights" in *Women's Lives, Men's Laws* (2005) Cambridge: Harvard University Press, page 151-161.

607 For information about the Swedish law regarding prostitution and trafficking, see Prostitutionresearch.com, Fact Sheet in Violence Against Women, http://www.prostitutionresearch.com/swedish.html.

608 The Swedish law on prostitution is a progressive law that decriminalizes prostitution for the victim only. Buyers, pimps, and traffickers are vigorously prosecuted by Sweden with felony-level charges against johns. Since the introduction of the law in 2000, trafficking from other countries into Sweden has plummeted. Traffickers obviously prefer to transport women to markets in countries where prostitution is legalized or tolerated. For more information on the Swedish law see http://www.prostitutionresearch.com/c-laws-about-prostitution.html.

609 From 2001 to 2006, Gunilla S. Ekberg was the Special Advisor to the Swedish government on prostitution and trafficking in human beings. She is currently Co-Executive Director of the Coalition Against Trafficking in Women International.

610 Gunilla S. Ekberg (2004) The Swedish Law that Prohibits the Purchase of Sexual Services. *Violence Against Women* 10:1187. Available at http://www.prostitutionresearch.com/c-laws-about-prostitution.html. Also available at http://action.web.ca/home/catw/readingroom.shtml?x=67517&AA_EX_Sess ion=799a0b96e5ec74f2368c124d87eb9db3. See also Gunilla S. Ekberg (2001)

Prostitution and Trafficking: the Legal Situation in Sweden. March 15, 2001. Unpublished manuscript on file with author.

611 Brown, *Sex Slaves,* here refers to families exploiting women as well as men buying women for sex because in India families sell girls into prostitution. At first glance, some readers might assume, "that's a Third World phenomenon." The author reminds readers of the case of a woman in legal Nevada prostitution who was pimped into a brothel by her parents. This is described in the chapter on legal brothel prostitution in this book.

612 Brown, *Sex Slaves,* page 200.

613 Antonella Gambotto-Burke (2007) Call Girls: Private Sex Workers in Australia / Not for Sale: The Return of the Global Slave Trade. Book review for *The Australian.* News.com.au Online. Retrieved July 5, 2007 from http://www.theaustralian.news.com.au/story/0,20867,21693483-5003900,00.html.

614 Personal Communication (2007) with Gunilla S. Ekberg. July 4, 2007.

615 AFTENPOSTEN (2007) Buying Sex Can Yield Jail Term. *News from Norway.* Retrieved July 5, 2007 from http://www.aftenposten.no/english/local/article1870915.ece.

616 For descriptions of the 2007 New York antitrafficking law, see http://www.stophumantraffickingny.org/release0606LAW.html and http://www.ny.gov/governor/press/0606071.html and http://www.criminaljustice.state.ny.us/legalservices/programbill_s5902.htm. For a copy of the 2007 New York antitrafficking law, contact Taina Bien-Aime, Executive Director, Equality Now at info@equalitynow.org or www.equalitynow.org. See also www.stophumantraffickingny.org.

Under New York's 2007 anti-trafficking legislation, sex trafficking is a Class B felony, which could entail up to 25 years in prison. Labor trafficking is a Class D felony, which could lead to jail time of up to 7 years. The New York antitrafficiking law increases the penalties on patronizing prostitution, which went from a B to an A misdemeanor, and clarifies the existing New York law on sex tourism.

The New York law provides trafficking survivors with a range of comprehensive services such as emergency housing, healthcare, drug addiction treatment, translation services, job training, and services related to immigration protection. It also requires law enforcement to coordinate with the federal government to assist victims in obtaining special visas that allow them to remain in the United States and eventually become eligible for refugee assistance; and

Finally, the 2007 New York antitrafficking law creates an interagency task force to recommend best practices for training and outreach to the law enforcement community and to service providers, as well as to gather data on the number of victims and effectiveness of the new law.

[617] Convention for the Suppression of the Traffic in Persons and of the Exploitation of the Prostitution of Others, 96 U.N.T.S. 271. See also Convention on the Elimination of All Forms of Discrimination Against Women, Dec. 18, 1979, art. 5, 1249 U.N.T.S. 13, 17, declaring in Article 6 that state parties shall take all appropriate measures, including legislation, to suppress all forms of traffic in women and exploitation or prostitution of women.

[618] See Janice G. Raymond (2002) The New U.N. Trafficking Protocol. *Women's Studies International Forum* 25: 491.

[619] Barb Brents and Kate Hausbeck, with COYOTE and Sex Workers Outreach Project are among the founders of Desiree Alliance which sponsored a conference at University of Nevada at Las Vegas in July 2006. Re-visioning Prostitution Policy: Creating Space for Sex Worker Rights and Challenging Criminalization organized sex workers and sex worker advocates and allies interested in working toward the goal of decriminalization of prostitution in the U.S. See http://www.desireealliance.org/conference.htm. See coverage of the conference at David Kihara (2006) Talkin' Dirty: Sex industry is topic at convention. *Las Vegas Review-Journal* July 14, 2006. Retrieved July 24, 2006 from www.reviewjournal.com/lvrj_home/   2006/Jul-14-Fri-2006/news/8487993.html.

[620] Neff, Legalized Prostitution.

[621] The first response of one man to the mention of a boycott was, "I just bought a home in Las Vegas. What about my mortgage?"

[622] Lisa Kim Bach (2006) Juvenile Prostitution: Trafficking in children on increase: Las Vegas among 14 U.S. cities where problem is most severe. *Las Vegas Review-Journal* March 19, 2006. Retrieved March 27, 2006 from http://www.reviewjournal.com/lvrj_home/2006/Mar-19-Sun-2006/news/6434154.html.

[623] Personal Communication (2006) with Candice Trummell Nye County Comissioner. July 21, 2006.

[624] Nevada Coalition Against Sex Trafficking is at www.nevadacoalition.org

[625] For a summary of the consequences of legal prostitution see Melissa Farley (2004) 'Bad for the Body, Bad for the Heart:' Prostitution Harms Women Even If Legalized or Decriminalized. *Violence Against Women* 10: 1087-1125. See also Janice Raymond, Ten Reasons for *Not* Legalizing Prostitution.

[626] Mary Sullivan and Sheila Jeffreys (2001) *Legalising Prostitution is Not the Answer: the Example of Victoria, Australia.* Coalition Against Trafficking in Women Australia and USA. Available at www.catwinternational.org

[627] Sullivan, *Making Sex Work*, page 332.

[628] Sullivan, *Making Sex Work*, page 334.

INDEX

About the Author

Melissa Farley, Ph.D. is a research and clinical psychologist who has practiced psychotherapy for 40 years. She spent the past 14 years conducting international research on prostitution, focusing on the psychological harms of prostitution and trafficking.

Farley is author of 25 peer-reviewed articles on the topics of sexual violence, prostitution, and trafficking. In 2003, she edited *Prostitution, Trafficking & Traumatic Stress* a collection of articles by 30 experts in the field published by Haworth Press. Farley's research has been used by governments in South Africa, New Zealand, Spain, Ghana, Korea, and the United States for education and policy development. She has been an associate scholar with the Center for World Indigenous Studies since 2002. (www.cwis.org). Melissa Farley has consistently addressed the roots of trafficking for prostitution and pornography in racism, sexism, and poverty. She emphasizes the overlap of domestic and international trafficking.

Dr. Farley is Director of Prostitution Research & Education (PRE) a nonprofit organization in San Francisco whose mission is to advocate for alternatives to prostitution and trafficking including policy and legal change, using research, public education, and the arts. The PRE website (prostitutionresearch.com) provides information about trafficking, prostitution, and pornography, receiving 50,000 page views per month from survivors seeking support or services, family members, policymakers, students and advocates. The website includes a list of agencies offering services to women who are escaping prostitution.

Melissa Farley is a member of the Nevada Coalition Against Sex Trafficking. See (www.nevadacoalition.org).